Critical Adult Education in Food Movements

Colin R. Anderson • Rosa Binimelis Adell •
Michel P. Pimbert • Marta Rivera Ferre
Editors

Critical Adult Education in Food Movements

Previously published in *Agriculture and Human Values* Volume 36, Issue 3, September 2019

Springer

Editors
Colin R. Anderson
The Institute for Agroecology
University of Vermont
Burlington, VT, USA

Michel P. Pimbert
Coventry University
Coventry, UK

Rosa Binimelis Adell
Arran de terra SCCL
Santa Maria d'Oló, Spain

Marta Rivera Ferre
INGENIO (CSIC-Universitat
Politècnica de València)
Valencia, Spain

ISBN 978-3-031-19399-6

This Springer imprint is published by the registered company Springer Nature Switzerland AG
The registered company address is: Gewerbestrasse 11, 6330 Cham, Switzerland

Contents

Agriculture and Human Values (2019) 36:521–529
https://doi.org/10.1007/s10460-019-09941-2

SYMPOSIUM/SPECIAL ISSUE

Introduction to the symposium on critical adult education in food movements: learning for transformation in and beyond food movements—the why, where, how and the what next?

C. R. Anderson[1] · R. Binimelis[2] · M. P. Pimbert[1] · M. G. Rivera-Ferre[2]

Accepted: 25 April 2019 / Published online: 15 May 2019

Many social movements, peasant organisations and radical educators have been deeply engaged in education for food system transformation for a relatively long time. In contrast, scholarly work on the possible role of pedagogy in radically transforming food systems is more recent. But this field of inquiry is growing as new insights for theory and practice are emerging—especially in the agroecology literature. This symposium of Agriculture and Human Values is timely because it focuses on new research on the importance of critical education for the spread of food sovereignty and agroecology to more people and places. It pays particular attention to the important role that learning, education and pedagogy can play in social transformation for food sovereignty and justice—a tradition that we refer to broadly as "Learning for Transformation". It draws together 7 articles that offer new critical insights about why, where, and how learning for transformation is being implemented,—and what next. In this editorial introduction, we reflect on how the different educational processes and methods presented here point to a range of possibilities and challenges for social movements.

The why: critical education in an era of multiple crises

Writing about the decline of reason in everyday life, Stanley Aronowitz highlighted 40 years ago the dangers of a citizenry that lacks the capacity for critical thought, "The issue is the capacity for theoretical or conceptual thought itself. When people lack such competence, social action that transcends the struggle for justice within the empirically given rules of social organisation and discourse is impossible… critical thinking is the fundamental precondition for an autonomous and self-motivated public or citizenry" (Aronowitz 1977). At heart, critical education is a process of cultivating consciousness for reason, action and social justice. From this perspective, being conscious 'is a radical form of being' (Freire 1978) in which education helps learners to understand the oppressive systems around them, to critically analyse their situation, and to link theory and action for positive change.

Education for critical consciousness, reflexive reasoning, ethical action, and justice is needed more than ever today in an era where people and planet are confronted with profound threats. Developing a critical awareness of the unfolding ecological, economic, social, political, and cultural crises is essential. But this is not enough in and by itself. We need a shift in thinking amongst all learners to better understand the root causes of these interrelated crisis, and not just see the more immediate reasons for a financial crisis, an ecological disaster, the rise of the far right and xenophobia, or farmer suicides. The articles in this symposium thus address what we refer to as a 'learning for transformation' which aims to construct a critical education and nuanced engagement with underlying causes to advance possibilities for deep change. This iterative process of learning provides the basis for collective action and social change.

A commitment to critical education is key for the future of food and agriculture and social transformation. Indeed, critical learning may be a prerequisite for deepening the transformative vision and praxis of social movements. For example, as the momentum for agroecology as an alternative paradigm for food and farming grows, so have the risks of co-optation and re-alignment with productivist, neoliberal and corporate-controlled farming (Anderson et al. 2019; Giraldo and Rosset 2018; Rivera-Ferre 2017; Pimbert

✉ C. R. Anderson
colin.anderson@coventry.ac.uk

1 Centre for Agroecology, Water and Resilience, Coventry University, Priory Street, Coventry CV1 5FB, UK

2 Agroecology and Food Systems Chair, Inclusive Societies, Policies and Communities Group, University of Vic-Central University of Catalonia, 08500 Vic, Spain

2018a). Critical education is, more than ever, necessary to understand how to stay rooted in a transformative perspective for food, agriculture, and human values. An education for critical consciousness can also help create linkages and solidarities between different social movements by fostering new dialogues that reach across boundaries in meaningful ways. Such an educational process may indeed be required to create synergies across movements struggling for food sovereignty (https://viacampesina.org), environmental justice (https://ejatlas.org), feminisms (https://womensmarchglobal.org/), climate justice (https://climatejusticeaction.net), degrowth (https://degrowth.org) or racial justice (https://blacklivesmatter.com)—for example.

The where: education for transformation in, between and beyond formal and informal spaces of learning

Where is *learning for transformation* occurring? Where is it situated? Learning for transformation is based on reflexive action that confronts and dismantles oppression. This contrasts with most approaches to education where learning is oriented towards understanding and adapting behavior within and according to the dominant system (Giroux 2013). Transformative education thus takes place in reference to, but also within and against, dominant power relations and structures that shape the nature of learning and education. The question of where such education processes take place is then not only a matter of cartesian geography (e.g. rural, urban, in Latin America, in Cairo). It is also contingent on relations and spatialities of power. *The where* of learning for transformation is defined in its relation to power—occurring within, beyond, in reaction to, as a counterpoint, at the margins of, or in the face of various systems that generate social and ecological injustices. It thus takes place wherever agency, power and social change are being claimed by the marginalised through processes of critical learning (Freire 1970; Hooks 1997).

Learning is often discussed in relation to a binary between formal and informal learning spaces or sectors—and research on adult education in food movements often focuses on one or the other. In the informal sector for example, peasant networks in Central America and Cuba (*Campesino a Campesino*), the landless workers movement in Brazil (MST), the European Coordination of La Via Campesina (ECVC), and indigenous groups in Mexico are systematizing learning programs autonomously from any formal institutions (Anderson et al. 2019; Barbosa 2016; Meek et al. 2019; McCune and Sánchez 2019). These initiatives and programs are often led by farmer- and citizen- educators. In many cases they have emerged to address the lack of appropriate opportunities in formal education and extension which mostly promote Green Revolution approaches and devalue local and traditional knowledges. In many cases, these grassroots initiatives are also an intentional political response against mainstream education which is viewed as deficient and an expression of imperialist hegemony. In contrast, education for transformation involves developing pedagogies that are rooted in the cosmovisions, interests, and political situations of the oppressed in order to build social movements and social change.

In the formal sector, several programs on agroecology, food sovereignty and social transformation do exist. To different degrees, these programs are situated more squarely within the dominant system, close to centres of intellectual, political and economic power. These programs have substantial resources. They are often enjoy wider credibility within the dominant system as institutions of 'higher learning' (note "the where" = above). However, there are few programs in the formal sector that are substantially embedded,—or even marginally connected to—, wider processes of social transformation based on agroecology and food sovereignty. Most university programs and their pedagogies are shaped by the wider political economy that values and emboldens neoliberal-commercialisable knowledges and elite knowledge systems, and mainly perpetuate and further entrench the status quo (People's Knowledge Collective 2017).

On the other hand, despite the many disincentives and pressures that undermine attempts to enable learning for transformation within formal education (Anderson 2019), there are impulses, actors and programs—often in the margins of academia—which are more closely aligned with social movements and the politics of collective transformation (see: Borras 2016; Levkoe et al. 2018; Méndez et al. 2015). These programs and people are guided by a commitment to *community engaged learning and research*, traditions of *participatory action research,* or more general commitments to the wellbeing of marginalised peoples in and beyond their locale. In this regard, programs like the university training program presented in this issue by López-García et al. (2019), provide examples of university anchored programs that can help to promote agroecological transitions in territories.

Formal programs are much more likely to reflect a learning for transformation when they work with social movements. Organisations promoting food sovereignty have in some cases intentionally engaged in adult learning collaborations with post-secondary educational institutions. Here, the intention is to centre knowledge from the margins of society in the institutions of elite knowledge production. Rebecca Tarlau's (2019) recent book demonstrates the important role that the Landless Farmers Movement (MST)'s work on educational reform in Brazil has played as one of multiple repertoires of action in a wider social change strategy. The work of the MST shows how social movements have the capacity to implement their own pedagogical programs inside the formal schooling system(s), even under contradictory and often hostile conditions.

However, mainstream approaches and attitudes to education have historically served to perpetuate and re-entrench the *status quo* along with its dominant ideology. Althusser (1984) described education as an ideological state apparatus, in this case preparing students to accept and conform with the dominant food regime. Much more work is needed to understand how power imbalances can be addressed when mainstream institutions—with their expert cultures, academic priorities and neoliberal economics and incentive structures (Anderson 2019; Pimbert 2018b)—partner with social movement organisations and actors.

A critical learning approach for social transformation transgresses the binary divide between informal-formal education. Different pedagogical methods and tools are used in diverse contexts. But these different pedagogical processes are united by a worldview and a political commitment that consider all spaces of interaction as sites of learning where pedagogy can be used and developed in an iterative and reflexive way. In the context of food and agriculture, this includes spaces of exchange (e.g. shops, markets), in families, schools, libraries, pubs and the multitude of spaces in everyday life. Incorporating this commitment to critical adult learning goes beyond the usual conceptions of lifelong learning as an individualistic and self-referential refrain. Instead, it positions critical learning as a collective and all encompassing mode of social learning, activism and change. In this sense, all spaces and all interactions have pedagogical value that can be intentionally amplified.

Learning for transformation is then made up of individual and collective agency which works across fluid, subterranean, rhizomatic processes that manifest concretely in planned moments as well as at times of spontaneous learning. Learning for transformation transgresses any fixed categorisation or prescription of where education should take place and of who is qualified to be a teacher in any place versus another. Once an individual or organisation sees themselves in this light, and views all moments as having pedagogical value (for teaching–learning), the division between formal and informal spaces of learning becomes less important and less constricting. The division between who is a teacher and who is a student also begins to dissolve, as shown in different schools in Latin America (Meek et al. 2019; McCune and Sánchez 2019) and Spain (Casado Baides 2018). In this formulation, the 'where' of learning for transformation is normatively, "everywhere" and by everyone.

Learning for transformation is a fundamentally subversive process in that it seeks to contribute towards just societal transformation. Education, in this regard, is deliberately and explicitly political: not an education about—but an education for—normative visions of change such as food sovereignty, social justice or emancipation. In terms of the question of "the where" then, it is important to remember that it is also a question of "where to". The focus on transformation recognises that where we are today is not where we should be in the future. Learning and pedagogy are viewed as important approaches to gaining collective agency and determining the pathway(s) of emancipation.

In this regard, not all food systems education reflects an approach that contributes to a learning for transformation. Indeed in many cases, education that is deemed to be 'progressive' or 'for food sovereignty' can inadvertently be apolitical, de-politicising or de-mobilising if it fails to develop the political consciousness and actions needed for social transformation. For example, many NGOs, governments and even social movement initiatives engage learners as individuals, focusing on proximate (rather than deep systemic) analysis of political problems. Or even more problematic, they construct learning for transformation as a process of only acquiring new skills as individual consumers. In some cases, it is possible that these actors have a radical sense of the "where to". But their understanding of "the how" in regards to learning and education either lacks depth and critical thought or the capacity and skills to follow a transformative path. As evident across the articles in this issue, learning for transformation requires a deeply political pedagogy that transcends individuals. It is concerned with the wider 'how' of emancipatory learning and collective processes of change. Indeed, a commonality across all of these contributions is an emphasis on documenting, developing, and theorizing "the how"—that is the pedagogical processes that underpin an education for food sovereignty and food systems transformation.

The how: pedagogical underpinnings of learning for transformation

The seven articles in this special issue highlight new thinking on how,—and under what conditions—, can pedagogies for food sovereignty and agroecology contribute to a community's political and physical control over food systems and the governance of their territories (Meek and Tarlau 2016). Many of the articles articulate and further elaborate some long standing principles of critical pedagogy, horizontal learning, transformative learning and other related traditions. We encourage readers to work through the articles to get a sense of how these approaches are being developed in different contexts and regions including Africa, North America, South America, Europe and Asia. In the following sections, we emphasise five important areas to consider when reading across the seven papers.

The political economy of education: a shift from a commercial and individualised entrepreneurial model of training to a commitment to education for solidarity and care

Many of the articles demonstrate how the political economy of education shapes the potential of a learning for transformation approach. Massicotte and Kelly-Bisson (2019) examination of permaculture training in Eastern Ontario (Canada) for example illustrates how the individualised entrepreneurial tendencies that predominate in permaculture training undermines the emancipatory potential that is ostensibly written into the permaculture approach. They argue that widespread access to permaculture knowledge and skills today is limited because it depends on payment for certified training courses led by expert trainers in a market-led approach to education. In this case, the wider political economic context of western capitalism and neoliberalism has shaped the delivery and scope of teaching and educational programs, thereby reducing its initial transformative potential. As a result, permaculture design trainers in Canada "have tended to function as technical trainers from affluent social backgrounds without an explicit intention of participating in a transformation of agri-food systems" (Massicotte and Kelly-Bisson 2019).

In another article in the special issue, McCune and Sánchez (2019) explain how the influence of western NGOs in a period of neoliberal development reshaped the Peasant-to-Peasant approach in Nicaragua away from one based on mutual-aid. A service-provision and market led approach emphasized paid training courses to teach peasants to become service providers through ecotourism and other commercial activities. Both Massicote and Kelly-Bisson as well as McCune and Sánchez draw a contrast between these kinds of individualist pedagogies with an education for solidarity. They use examples from Brazil and Cuba to emphasise the potential of an educational model based on alternative political economic relations and imaginaries. They reject capitalist values and modes of educational production as the basis of meeting the material and intellectual needs of instructors and activists-learners-farmers. The enactment of prefigurative post-capitalist forms of exchange and mutual support creates new possibilities that are otherwise systemically suppressed.

Meek et al. (2019) discuss the issue of dependence of grassroots educational programs on external funding, and how this undermines more radical pedagogies. They ask to what extent can funding, institutionalization (or the institutional environment) and/or the relation with the state compromise the emancipatory nature of these experiences. This is a common situation in social movements where the absence of funds often stimulates creativity and collective work. In contrast, the availability of substantial funds can reduce levels of grassroots voluntary work and initiatives by hiring external technical staff. This can subsequently undermine collective work, unless the recruited staff are well trained in facilitating cooperative group processes based on clear commitments to change. As Meek et al. (2019) state, "institutionalization can be the death knell for critical food systems education", where transformative aims are eroded or sidelined through funding and reformist actions supported by the state, philanthropic donors, or other actors aligned with the dominant regime.

The decommodification of education—and resisting commodification in the first place—can open new possibilities for solidarity-based and politicized learning programs that are accessible to all,—including indigenous peoples, women, as well as other subaltern and oppressed groups. Embedding this solidarity in social movements in ways that strengthen organisations, networks, radical visions and alternative identities (Anderson et al. 2019) helps sustain the practices in which a solidarity economy of education can grow.

A critical pedagogy that values organic intellectuals

Contributors to this special issue affirm the importance of organic intellectuals and educator-trainers in enabling mutual learning for political organizing and knowledge-building. According to Gramsci, organic intellectuals emerge organically from a particular social-political class and play a critical role as facilitator-organisers of counter-hegemonic struggles. They often act as educators, helping to construct and translate counter-hegemonic theory into educational praxis to nurture class consciousness in social movements (Ramos 1982). The role of organic intellectuals as educator is essential to achieve unity of theory and practice. By linking the abstract and the concrete, organic intellectuals and educators can foster a counter-hegemonic movement capable of contesting the dominant agri-food regime. In this regard, the articles reveal two critical components of a learning for transformation: (i) every person has the capacity to act as an 'intellectual' and indeed to foster and realise this capacity; and (ii) it is from the oppressed, and through the leadership of organic intellectuals, that transformative knowledges can emerge.

These two elements are linked and can overlap, but we differentiate them because they have different outcomes in education for food sovereignty. The first one favours agroecological education because it implies that facilitator-educators as organic intellectuals need to be humble, avoid vanguardist tendencies, and strengthen collective forms of intelligence through horizontal approaches to learning. This approach is rooted in a recognition that counter-hegemonic leadership is strongest when distributed, when intellectual power is collectivized, and when the inherent intellectual capacities of the oppressed are cultivated equally.

The second important characteristic is that it is from "the oppressed" people that knowledge for social transformation can emerge. Gramsci believed that an emancipatory project should emerge from marginalised communities becoming conscious of the multiple oppressions they may suffer, and by organising themselves (Massicotte and Kelly-Bisson 2019). This is in fact how political alternatives like food sovereignty were born. It is indeed from marginal groups that new knowledges may emerge in future. This view also coincides with feminist theory which argues that the marginalised and oppressed have the knowledge and experience needed to understand and confront their own problems (Truman et al. 2005; Brown and Strega 2015).

Gramsci's concept of "organic intellectuals" is key in this process of fundamental change. Organic intellectuals emerge from within a particular class, —middle or working class,—they are recognised by the people of this class, and they take on the role of "specialized" intellectuals capable of fostering a sense of self-awareness among members of their class. Typically, organic intellectuals work to mobilise and politicise within their respective communities to catalyse, facilitate, and support deep cultural and socio-political change,—from the bottom-up.

It is noteworthy that organic intellectuals and other facilitator-trainers must not only have the appropriate knowledge and skills to train others. Organic intellectuals and other educators for food sovereignty also need to cultivate enabling attitudes and human qualities such as humility and empathy to work with the most marginalised sectors of society. Indeed, McCune and Sánchez (2019) argue that, "Human qualities such as humility, honesty, integrity and solidarity are considered as important to the learning process as are composts, intercropping, and seed saving".

An emphasis on collective learning and cooperation

Learning for transformation is not an individualistic endeavour. It is based on an intentional collectivity in organisations, programs, and affinity groups as well as in emergent networks of people linked through bonds of reciprocity, mutual aid, and communication. Collective processes, such as horizontal farmer-to-farmer exchanges, *diálogos de saberes,* and cooperation allow agroecological knowledge to be shared, documented, discussed, built and mobilized on a large scale, connecting many people and places (Anderson et al. 2019; McCune and Sánchez 2019; Mann 2018). These processes recognise that when learning is conceived as an individual endeavour, this often gives rise to individualistic subjectivities and parochial self-interested understandings. In turn, this undermines possibilities for transformation.

Collective learning dynamics fundamentally challenge the conventional view that "learning is an individual process that takes place in one human mind at a time" (McCune and Sánchez 2019). Anderson et al. (2019) argue that collective processes of learning are themselves an act of resistance against the individualising tendencies of the dominant neoliberal mode, and are thus critical for societal transformation. In this regard, it is often the informal learning spaces,—embedded in communities, networks and social movements,—where collective learning is most viable. This is largely because the learning dynamic is rooted outside the individualising nature of 'individual excellence', individual grading and individualised systems of meritocracy found in schools, universities and many workplaces.

The process of building enabling educational environments that adopt a collectivist mindset is therefore extremely important in transformative education (Casado Baides 2018). The educational environment involves intentionally structuring learning where students and learners are mutually a part of,—and responsible for,—building the community. This includes taking care of each other, creating a governance structure and rules for decision-making, as well as developing conviviality and a shared culture within the learning environment. This approach, however, takes time and often requires negotiating differences in a group. While challenging, these 'edges' between people, perspectives and worldviews create the conditions for deep learning that extends beyond the boundaries determined by a more individualistic and self-centred learning approach. Articles of this volume (McCune and Sánchez 2019; Meek et al. 2019) highlight the importance of the educational environment, in which students are part and responsible of the process of building community.

A focus on intersectionality

Overcoming the enduring and interconnected systems of economic, racial, colonial and gender oppressions,—and the violence that maintains them,—is another major challenge for emancipatory pedagogies. The design of learning for agroecology and food sovereignty is often based on an analytic framework that tries to identify how interlocking systems of power impact those who are most marginalized in society. A pedagogical praxis that is sensitive to issues of intersectionality works to address the complexities of power and inequality through a complex understanding of how people interact in learning spaces at the intersection of multiple axes of difference (e.g. gender, class, race, age, religion, sexual orientation, and disability). Pedagogical approaches that impose pre-existing and universal framings of oppression and political struggle can alienate learners whose experiences reflect distinct trajectories, identities and positionalities. An intersectional approach can create important opportunities for learning, solidarity and change across these diverse experiences and identities.

Bezner Kerr et al. (2019) describe how a focus on intersectionality and social inclusion guided the development of an innovative curriculum for agroecology in Malawi and Tanzania. Aimed at smallholders, the learning process was participatory and built on popular education, feminist praxis, experiential-based pedagogies, and theatre. As this critical pedagogical process strived to coherently weave together technical components (agroecology, climate change, soil nutrition….) it also had to sensitively engage with clashes of language, cultural norms, and terminology that reflected differences in gender and other axes of difference (Bezner Kerr et al. 2019). In another example from Canada, the excluded voices of migrant hired farm workers were foregrounded through a praxis of community-based arts (Perry 2019), primarily rooted in Augusto Boal's techniques of collective theatre creation (Boal 1985). This allowed participants to express and discuss their experiences of oppression. Plays created and performed by migrant farm workers made it possible to explore difficult issues related to intersectionality, such as deeply uneven power relations in the workplace, co-worker harassment, and loneliness.

By focusing on various intersections of social inequality as the matrix of domination (Collins 1990), critical pedagogical processes and methodologies can thus help de-stabilise and overcome interlocking vectors of oppression and hegemonic power in food and agriculture.

Regenerating territories through popular education

Large-scale programs of learning for transformation are most impactful when they embed collective learning in wider processes of territorialisation within radical frameworks such as agroecology and food sovereignty—or broader mobilising concepts like *Buen Vivir*. Thus, a process of territorialisation seeks not only to regenerate sustainable farming and agroecosystems and other parts of the food system (e.g. farmer seed networks, mills, dairies, community food processing facilities…) but also the territorial relations and infrastructure required to meet human needs and generate local livelihoods (e.g. schools, health clinics, community centers, cooperatives, small scale industry…). Learning for transformation thus knits together the human-ecological relations within territories that have often been stripped of their cultures, people, resources, and autonomy through centuries of capitalism and colonialism (Massicotte and Kelly-Bisson 2019).

Learning for transformation as a process of re-territorialisation is simultaneously immaterial and material (Giraldo and Rosset 2018). It involves learning cycles of both action and theory that simultaneously work to challenge ongoing processes of capitalist de-territorialisation whilst reflexively constructing alternatives. For example, learning programs can unpack and explore immaterial notions of solidarity in economic exchange. In so doing, they can re-invent new forms of modernity and well-being within their own specific territories while supporting the material project of developing alternative systems of economic exchange to exit unfair commodity markets (Pimbert 2018c). Thus, McCune and Sánchez (2019) show how, "As pedagogical processes develop, new territorial dynamics take shape, revealing that self-organized peasant education contributes to sustaining popular territories against the aggression and violence of globalized capital." Processes of "territorial learning" strengthen local organisations and wider social movements (Anderson et al. 2019), thereby enabling learners to become transformative agents of their own reality. López-García et al. (2019) describe a process of "agroecological dynamisation" that reflects this territorial approach, arguing that universities, under the right conditions and commitments, can play a role in mobilizing a wide diversity of actors (educators, farmers, consumers, policy-makers, etc.) to strengthen the social fabric in a territory.

The what next: new frontiers for education for and beyond food sovereignty

The papers in this issue, and our collective process of putting together the special issue as a team, have surfaced several unresolved questions and potential frontiers for *learning for transformation*. We share these here to prompt further thought on the "what next" for extending, deepening, and strengthening pedagogic praxis for transformative learning in struggles for food sovereignty and social justice.

Food sovereignty is, in theory, an emancipatory approach to framing and pursuing societal transformations. Yet, in many cases, the networks and practices of food sovereignty, agroecology, local food, permaculture, reproduce inequities and power differentials (People's Knowledge Collective 2017). Whilst there have been efforts to advance a theory of education for food sovereignty, including in this special issue, one of the yet to be explored frontiers is the extent to which popular education initiatives in different contexts are advancing emancipation for the most marginalised and oppressed. Recent work on the links between gender and agroecology in Brazil's landless farmers movement (Schwendler and Thompson 2017), indigenous education (Goulet and Goulet 2014), and decolonial pedagogy (Wane and Todd 2018) point to further research needed to deepen our understanding of the politics of learning for transformation.

We also found ourselves reflecting on the notion of a *diálogo de saberes* ("dialogue of knowledges") and to what extent this dialogue can effectively take place between actors who believe in food sovereignty or other radical causes on the left with those that have other, and in some cases deeply conflicting politics (neoliberals, far-right). We desperately need to develop approaches that go beyond learning amongst like-minded and politically aligned actors. Bringing together individuals and groups with different views and positionalities can help to better understand difference, identify common experience, develop empathy, transform conflict, and unearth the roots of oppression. These dynamics are not often visible on the surface but can be deduced and strengthened through deep and critical collective learning. Is it possible that such learning processes could help raise critical consciousness and acceptance of difference? Can they help avoid authoritarian tendencies in populist movements struggling for food sovereignty, 'radical political agroecologies' (Cadieux et al. 2019) or degrowth (D'Alisa et al. 2014)? Further work is needed to understand how pedagogy can help to cross these deepening divisions in society.

We also found ourselves asking how critical education helps enable and politicize a process of broader awakening and organising for social transformation? For example, the concept of "quiet sustainability" or "quiet food sovereignty" has been used to express a politics of sustainability that is performed through the enactment of sustainable lifestyles by actors that are not intentionally nor overtly politicized (Visser et al. 2015; Kneafsey et al. 2017). While these practices may enact sustainable behaviors and practices, they do nothing to reveal and address the underlying systems of oppression that are left intact and unquestioned. They also tend towards individualistic tendencies that run counter to the collectivist nature of learning for transformation articulated in this special issue. Finally, arguing for such passive approaches to sustainability could be said to come from a place of privilege (who has the privilege to make lifestyle choices?). To us, learning for transformation is a direct challenge to such de-politisised subjectivities. Encouraging a wider critical awakening is urgently needed today given the violence, inequality and potential ecological collapse in territories as well as possibly larger and even planetary scales. More work is needed to understand how learning for transformation can reach beyond the already-politicised,—to meet people where they are (e.g. peasants and consumers practicing what is referred to as quiet sustainability), and draw people together in collective processes of learning tied to the territorial and global processes of social movements.

Answering these questions may require crossing boundaries and consciously hybridising the insights and practices of other traditions of emancipatory education outside and beyond food sovereignty. For example, much can be learnt from radical, critical, indigenous, feminist and anarchist pedagogies for collective learning and action that address internal power dynamics of movements. Radical critical pedagogy (Mclaren 1997) and participatory action research (Wakeford and Rodríguez 2018) offer ways of moving food sovereignty/justice into a more reflexive space.

Similarly, ideas and practices traditionally associated with social anarchism have repeatedly emphasised the importance of education for social transformation. Exploring the neglected traditions of anarchist education in Spain, France, UK and the USA, Suissa (2006) shows the extent to which social anarchists are committed to a radical and substantive educational process based on clear moral principles. Anarchists affirm the right of individuals to be "active agents creating the possibilities for their own future" (McKenna 2001, p. 52). In turn, this calls for a deeply transformative education based on freedom for creative experimentation, critical thought, and active problem solving. As Bookchin (1990, p. 189) notes:

> Sensibility, ethics, ways of building reality, and selfhood have to be changed through educational means, by a politics of reasoned discourse, experimentation and the expectation of repeated failures from which we have to learn, if humanity is to achieve the self-consciousness it needs to fully engage in self-management.

More generally, the new frontiers for food sovereignty and beyond will need to prioritise an 'education for radical democracy' (Amsler 2015) with its wealth of critical tools and practical methods. By emphasising a politics of possibilities, these pedagogies of hope can help cultivate counter-capitalism, resistance to domination, a non hierarchical sensibility, practices of freedom, as well as radical alternatives grounded in environmental and social justice.

Conclusion

In this editorial overview, we introduced the concept of learning for transformation, unpacking the seven articles in the special issue to make sense of the why, where (and where to), how, and what next? While these have been useful for making sense of the collective contribution of these articles and how they are situated in the literature, we assert that it is important also for practitioners to ask themselves these questions of their own pedagogical practice and programs. The responses to these questions in particular places and situations are context specific and will evolve over time. By asking ourselves these questions, and debating the nature and dynamics of our learning and education programs, we can iteratively develop a pedagogy that is simultaneously more radical, transformative, and more effective. To this end, we invite you to connect with us and with each other as critical friends with a mutual commitment to learning for a more just and sustainable world.[1]

[1] Two spaces convened by two of the authors (Pimbert and Anderson) offer opportunities to connect with a community of practice/ praxis working to develop a space of collective reflexivity and learn-

References

Althusser, L. 1984. *Ideology and ideological state apparatus. In essays on ideology*. London: Verso.

Amsler, S.S. 2015. *The education of radical democracy*. London: Routledge.

Anderson, C.R. 2019. Becoming a public scholar in the professional academy: Confronting the institutional, interpersonal and internalized dimensions of (re)professionalisation. *ACME: An International Journal for Critical Geographers*.

Anderson, C.R., C. Maughan, and M.P. Pimbert. 2019. Transformative agroecology learning in Europe: building consciousness, skills and collective capacity for food sovereignty. *Agriculture and Human Values*. https://doi.org/10.1007/s10460-018-9894-0.

Anderson, C.R., J. Bruil, M.P. Pimbert, J. Chappell, C. Kiss, and J. Milgroom. Forthcoming. *Agroecology now! The battle for the future of food*.

Aronowitz, S. 1977. Mass culture and the eclipse of reason. The implications for pedagogy. *College English* 38: 768–774.

Barbosa, L.P. 2016. Educação do Campo [education for and by the countryside] as a political project in the context of the struggle for land in Brazil. *The Journal of Peasant Studies* 44 (1): 118–143.

Bezner Kerr, R., S.L. Young, C. Young, M.V. Santoso, M. Magalasi, M. Entz, E. Lupafya, L. Dakishoni, V. Morrone, D. Wolfe, and S.S. Snapp. 2019. Farming for change: developing a participatory curriculum on agroecology, nutrition, climate change and social equity in Malawi and Tanzania. *Agriculture and Human Values*. https://doi.org/10.1007/s10460-018-09906-x.

Boal, A. 1985. *Theatre of the oppressed*. New York: Theatre Communications Group.

Bookchin, M. 1990. *Remaking society. Pathways to a green future*. Boston, MA: South End Press.

Borras, S.M. 2016. *Land politics, agrarian movements and scholar-activism*. Inaugural Lecture, 14. International Institute of Social Studies. https://www.tni.org/files/publication-downloads/borras_inaugural_lecture_14_april_2016_final_formatted_pdf_for_printing.pdf.

Brown, L., and S. Strega (eds.). 2015. *Research as resistance, 2e: Revisiting critical, indigenous, and anti-oppressive approaches*, 43–64. Toronto: Canadian Scholars' Press.

Cadieux, K.V., S. Carpenter, A. Liebman, R. Blumberg, and B. Upadhyay. 2019. Reparation ecologies: Regimes of repair in populist agroecology. *Annals of the American Association of Geographers* 109 (2): 644–660. https://doi.org/10.1080/24694452.2018.1527680.

Casado Baides, B. 2018. *Procesos de formación de campesinos y disputa territorial para construir soberanía alimentaria. Análisis de experiencias impulsadas por organizaciones de La Vía Campesina en Brasil y País Vasco*. PhD Thesis. University of Basque Country.

Collins, P.H. 1990. *Black Feminist thought: Knowledge, consciousness, and the politics of empowerment*. Boston: Unwin Hyman.

D'Alisa, G., F. Demaria, and G. Kallis. 2014. *Degrowth: A vocabulary for a new era*. London: Routledge.

Freire, P. 1970. *Pedagogy of the oppressed*. Harmondsworth: Penguin.

Freire, P. 1978. *Pedagogy in process: The letters to Guinea-Bissau*. New York, London: Continuum.

Footnote 1 (continued)

ing as critical friends. www.peoplesknowledge.org and www.agroecologynow.com.

Giraldo, O.F., and P.M. Rosset. 2018. Agroecology as a territory in dispute: Between institutionality and social movements. *The Journal of Peasant Studies* 45 (3): 545–564.

Giroux, H.A. 2013. *On critical pedagogy*. New York and London: Bloomsbury Academic.

Goulet, L.M., and K.N. Goulet. 2014. *Teaching each other: Nehinuw concepts and Indigenous pedagogies*. Vancouver: UBC Press.

Hooks, B. 1997. *Teaching to transgress*. London: Routledge.

Kneafsey, M., L. Owen, E. Bos, K. Broughton, and M. Lennartsson. 2017. Capacity building for food justice in England: The contribution of charity-led community food initiatives. *Local Environment* 22 (5): 621–634.

Levkoe, C.Z., J. Brem-Wilson, and C.R. Anderson. 2018. People, power, change: Three pillars of a food sovereignty research praxis. *The Journal of Peasant Studies*. https://doi.org/10.1080/03066150.2018.1512488.

López-García, D., L. Calvet-Mir, M. Di Masso, and J.L. Espluga. 2019. Multi-actor networks and innovation niches: university training for local Agroecological Dynamization. *Agriculture and Human Values*. https://doi.org/10.1007/s10460-018-9863-7.

Mann, A. 2018. Education for food sovereignty as transformative ethical practice. *Policy Futures in Education*. https://doi.org/10.1177/1478210318816251.

Massicotte, M.J., and C. Kelly-Bisson. 2019. What's wrong with permaculture design courses? Brazilian lessons for agroecological movement-building in Canada. *Agriculture and Human Values*. https://doi.org/10.1007/s10460-018-9870-8.

McCune, N., and M. Sánchez. 2019. Teaching the territory: agroecological pedagogy and popular movements. *Agriculture and Human Values*. https://doi.org/10.1007/s10460-018-9853-9.

McKenna, E. 2001. *The task of Utopia. A pragmatist and feminist perspective*. Maryland: Rowman and Littlefield Inc.

McLaren, P. 1997. *Revolutionary multiculturalism*. New York: Routledge.

Meek, D., and R. Tarlau. 2016. Critical food systems education (CFSE): Educating for food sovereignty. *Agroecology and Sustainable Food Systems* 40 (3): 237–260.

Meek, D., K. Bradley, B. Ferguson, L. Hoey, H. Morales, P. Rosset, and R. Tarlau. 2019. Food sovereignty education across the Americas: multiple origins, converging movements. *Agriculture and Human Values*. https://doi.org/10.1007/s10460-017-9780-1.

Méndez, V.E., C.M. Bacon, R. Cohen, and S.R. Gliessman. 2015. *Agroecology: A transdisciplinary, participatory and action-oriented approach*. Boca Ratón, London and New York: CRC Press.

People's Knowledge Editorial Collective (Eds). 2017. *Everyday Experts: How people's knowledge can transform the food system*. Reclaiming Diversity and Citizenship Series. Coventry: Coventry University. http://www.coventry.ac.uk/everyday-experts.

Perry, J.A. 2019. Images of work, images of defiance: engaging migrant farm worker voice through community-based arts. *Agriculture and Human Values*. https://doi.org/10.1007/s10460-018-9861-9.

Pimbert, M.P. 2018a. Global status of agroecology, a perspective on current practices, potential and challenges. *Economic and Political Weekly* 53 (41): 52–57.

Pimbert, M.P. 2018b. Democratizing knowledge and ways of knowing for food sovereignty, agroecology and biocultural diversity. In *Food sovereignty, agroecology and biocultural diversity. Constructing and contesting knowledge*, ed. M.P. Pimbert, 259–321. Londo: Routledge.

Pimbert, M.P. 2018c. Food Sovereignty and the regeneration of terraced landscapes. *Annales, Series Historia and Sociologia* 28 (4): 779–794.

Ramos Jr., V. 1982. The concepts of ideology, hegemony, and organic Intellectuals in Gramsci's Marxism. *Theoretical Review* 30: 34.

Rivera-Ferre, M.G. 2017. The resignification process of agroecology: Competing narratives from governments, civil society and

intergovernmental organizations. *Agroecology and Sustainable Food Systems* 42(6): 666–685.
Schwendler, S.F., and L.A. Thompson. 2017. An education in gender and agroecology in Brazil's Landless Rural Workers' Movement. *Gender and Education* 29 (1): 100–114.
Suissa, J. 2006. *Anarchism and education. A philosophical perspective*. London: Routledge.
Tarlau, R. 2019. *Occupying schools, occupying land: How the landless workers movement transformed Brazilian education*. Oxford: Oxford University Press.
Truman, C., D.M. Mertens, and B. Humphries. 2005. *Research and inequality*. London: UCL Press, Taylor & Francis Group.
Visser, O., N. Mamonova, M. Spoor, and A. Nikulin. 2015. 'Quiet food sovereignty' as food sovereignty without a movement? Insights from post-socialist Russia. *Globalizations* 12 (4): 513–528.
Wakeford, T., and J. Sánchez Rodríguez. 2018. Participatory action research: towards a more fruitful knowledge' In *Connected Communities Foundation* Series, ed. Facer, K. and K. Dunleavy. Bristol: University of Bristol/AHRC Connected Communities Programme. https://connected-communities.org/wp-content/uploads/2018/07/Action_Research_SP.pdf.
Wane, N.N., and K.L. Todd (eds.). 2018. *Decolonial pedagogy: Examining sites of resistance, resurgence, and renewal*. Cham: Springer.

C. R. Anderson is a Senior Research Fellow at the Centre for Agroecology, Water and Resilience, Coventry University. His research focuses on community- and people- led processes of transformation for resilience, social justice and well-being. He works with communities, networks and organizations who are addressing the multitude of social, economic and cultural problems we face today and who are reimaging society and building alternatives. His collaborative research contributes to the development of thinking and action related to food systems, community building, food sovereignty, activism, and knowledge mobilization, and especially focuses on the workings of power, resilience, and regeneration in these areas. Visit www.agroecologynow.com and www.peoplesknowledge.org to find out more.

R. Binimelis holds a PhD in Environmental Science (Ecological Economics). She works as a researcher at the Chair in Agroecology and Food Systems and the research group "Societies, Policies and Inclusive Communities" at the University of Vic. Her research fields deal with the analysis, from a multidisciplinary perspective, of agri-food systems and socio-environmental conflicts. She is specialized in analysis of socio-economic impacts of GMOs and is interested in examining the social, cultural, political, ecological and economic implications of the ways food is perceived, produced and consumed. She has also worked on food sovereignty and gender, agroecology and alternative food systems. Rosa is now coordinating a network for food sovereignty in Central Catalonia (XaSACC).

M. P. Pimbert is Professor of Agroecology and Food Politics at Coventry University and the Director of the Centre for Agroecology, Water and Resilience in the UK. His research interests include agroecology and food sovereignty; the political ecology of biodiversity and natural resource management; participatory action research methodologies; and deliberative democratic processes. He works with networks of small and family farmers, indigenous peoples, and communities to advance transdisciplinary and transformative ways of knowing that regenerate local ecologies, economies and cultural diversity.

M. G. Rivera-Ferre is a Ramón y Cajal Research Fellow at the University of Vic-Central University of Catalonia. She coordinates the research line "Sustainable communities, social innovations and territories" within the group "Inclusive societies, policies and communities" and is director of the Chair Agroecology and Food Systems for Social Transformation. With a multidisciplinary profile in the analysis of social and ecological interactions of agri-food systems, she has a particular interest in alternative agri-food networks following the food sovereignty paradigm and in the analysis of feminists and commons theories as to be adopted in agri-food research. She has worked on the potential of local traditional agri-food knowledge in transformative adaptation to climate change. Marta has participated as lead author in the International Assessment of Agricultural Knowledge Science and Technology for Development (IAASTD) and the Intergovernmental Panel on Climate Change (IPCC).

Agriculture and Human Values (2019) 36:531–547
https://doi.org/10.1007/s10460-018-9894-0

SYMPOSIUM/SPECIAL ISSUE

Transformative agroecology learning in Europe: building consciousness, skills and collective capacity for food sovereignty

Colin R. Anderson[1] · Chris Maughan[1] · Michel P. Pimbert[1]

Accepted: 19 October 2018 / Published online: 12 November 2018

Abstract

Agroecology has been proposed as a key building block for food sovereignty. This article examines the meaning, practices and potentials of '*transformative agroecology learning*' as a collective strategy for food system transformation. Our study is based on our qualitative and action research with the European Coordination of Via Campesina to develop the European Agroecology Knowledge Exchange Network (EAKEN). This network is linked to the global network of La Via Campesina and builds on the strong experiences and traditions of popular education in Latin American peasant movements. Rather than focusing on agroecology education as a process of individual learning, we analyse how a transformative agroecology education can be strengthened as a critical repertoire of action used by social movements to advance food sovereignty. Our analysis contributes a new theory of transformative agroecology learning based on four key characteristics or qualities: horizontalism; *diálogo de saberes* (wisdom dialogues); combining practical and political knowledge; and building social movement networks. While these different elements of transformative agroecology learning were present across EAKEN, they were unevenly developed and, in many cases, not systematized. The framework can help to strategically and reflexively systematize and strengthen a transformative agroecology learning approach as a key building block for food sovereignty.

Keywords Agroecology · Pedagogy · Learning · Sustainable agriculture · Food sovereignty · Europe · Popular education

Introduction

Food sovereignty has emerged as the most important global discourse around which social movements are organising to contest and challenge neoliberal development (Wittman 2011). Food sovereignty has been taken up by rural and urban social movements around the world as a radical political project of food system transformation that emphasises the democratisation of agriculture and food, rights of food producers and autonomy from powerful external actors (e.g. agribusiness, political elite). As a transformative project, food sovereignty involves the creative re-visioning and fundamental re-design of whole systems. It requires *seeing things differently* from *business as usual* industrial food and farming, *doing better things* and re-thinking food systems on a participative basis (Pimbert 2008). This involves a questioning of the logics and accounting that underpin the dominant modes of agriculture through learning processes that force ruptures in conventional thinking and practice. This deep learning asks 'why' questions about underlying paradigms, norms and values as well as the governance and decision-making processes that frame and legitimize the purpose of knowledge, policies, organizations, technologies and practice.

Agroecology has been developed by social movements as an inseparable component of food sovereignty and a key component of its transformative project. Agroecology provides an alternative set of principles to organize the food systems based on maximizing the positive interrelations between people, farming and nature, and increasing the autonomy of farmers (Nyéléni 2015; Pimbert 2015; Wittman 2011). While work on food sovereignty tends to emphasize contentious political activities such as protest and advocacy, agroecology represents the on the ground articulation of food sovereignty in the practices of food producers as the basis for an alternative food system.

The alignment of agroecology and food sovereignty, however, is not always self-evident, largely because agroecology has become a concept with multiple meanings—there are

✉ Colin R. Anderson
colin.anderson@coventry.ac.uk

1 Centre for Agroecology, Water and Resilience, Coventry University, Priory Street, Coventry CV1 5FB, UK

multiple 'agroecologies' (Holt-Giménez and Altieri 2016). While agroecology can be rooted in food sovereignty, it can also be used with a narrow technocratic framing by mainstream institutions and scientists in ways that undermine its radical political potential (Holt-Giménez and Altieri 2016; Nyéléni 2015; Levidow et al. 2014; Pimbert 2018; Levidow 2015). On the other hand, social movements are developing and employing strategies to better link agroecology and food sovereignty. However, there is little known about the specific processes that social movements are using to ensure that agroecology and food sovereignty are mutually reinforcing.

In this paper, our goal is to examine the key principles guiding the emergence and implementation of a transformative agroecological learning approach. Learning and knowledge exchange, which have been endorsed both by social movements (Nyéléni 2015) and mainstream institutions (FAO 2018; AgriCultures 2013), are a key area to support the amplification or the scaling of agroecology. Following Mier y Terán et al. (2018), we see such scaling as inclusive of different actors with vastly different politics and intentions who co-define and shape agroecological knowledge, learning and innovation. In this context, different actors with different politics and intentions are required to define and shape agroecological knowledge, learning and innovation (IAASTD 2009; Levidow et al. 2014; Pimbert 2018). Thus, it is critical to examine and articulate the collective processes of learning being advanced by its practitioners. Negotiated and developed from the grassroots level, we argue that agroecological learning helps to weave the connective tissue between the more abstract politics of food sovereignty and the in situ practice of agroecology.

This article draws from our work with the European Coordination of La Via Campesina (LVC) to establish the European Agroecology Knowledge Exchange Network (EAKEN). Through a qualitative and action research approach, we develop a framework that characterises transformative agroecological learning for food sovereignty. This paper thus contributes to the growing literature on the strategies and challenges of implementing food sovereignty in different contexts (Edelman 2014) and on agroecology as an alternative development paradigm (Pimbert 2015; Levidow et al. 2014; Martínez-Torres and Rosset 2014). The article also fills a geographic gap, focusing on agroecology and agroecology learning in European social movements. Whereas an "agroecological revolution" has ostensibly taken hold in some parts of Latin America (Altieri and Toledo 2011), agroecology is a relatively new concept in Europe and there is a need for critical inquiry into the dynamics, tensions and opportunities at play there.

Agroecology and food sovereignty

Food sovereignty first appeared as a prominent discourse in the public sphere through La Via Campesina at the 1996 World Food Summit in Rome (Patel 2009). It arose in the context of growing rural crises stemming from political and structural conditions, including the dismantling of infrastructural support, the liberalization of agricultural markets, and the increasing enclosure of food and agricultural policy-making by transnational entities such as the WTO (Desmarais 2007; Martínez-Torres and Rosset 2014; McMichael 2008).

Food sovereignty emphasizes the democratisation of food systems, policy, practice and knowledge, particularly the rights and autonomy of food producers (Nyéléni declaration 2007; Wittman 2011; Pimbert 2008). It also entails a rejection of the politics that have come to underpin the dominant food security discourse, which is viewed as being based on technical fixes rather than addressing the political underpinnings of problems in the food system (Lang 2010). Thus, food sovereignty in many ways arose to contest the dominance of scientists, NGOs, policy makers and corporates as the sole protagonists in the development of food systems.

Despite substantial popularity among social movement groups, food sovereignty has not escaped controversy within academic circles. Criticisms have clustered around a perceived 'lack of specificity' (Edelman 2014; Patel 2009; Aerni 2011), its vulnerability to corporate capture or containment (Alonso-Fradejas et al. 2015), fears regarding its overlap with crude nationalism or 'agrarian populism' (Bernstein 2014), as well as limited applicability outside rural space and the Global South (Aerni 2011). In this context, agroecology has been advanced by some as a counterpoint to food sovereignty's more abstract political objectives, that is, not only as a science and a set of agricultural practices, but also as a repertoire of practices capable of resisting corporate capture and containment (Pimbert 2008; Altieri 2009; Wezel et al. 2009).

Increasingly, social movements have positioned the food sovereignty project as inseparable from agroecology as a paradigm for an alternative food system (La Via Campesina 2013; Nyéléni 2015). Agroecology promotes a functional biodiversity and nutrient cycling and the principle of working with nature (Altieri 2009; Gliessman 2014). It seeks to maximize the autonomy of food providers, emphasising and drawing on farmers' knowledge and local resources, while challenging power structures for social and ecological transformation (Nyéléni 2015; Pimbert 2015; Wezel et al. 2009). Rather than imposing a prescriptive framework, agroecology has been praised for its articulation of a set of values and principles flexible enough to allow bespoke implementation in specific territorial contexts, reflecting the social, political and biocultural contingencies of place (Nyéléni 2015; Francis et al. 2013; Méndez et al. 2015). This sharply contrasts with top-down attempts to replicate a generalised agroecology which immediately divorces it from the place-based, social and political moorings which define agroecology (González de Molina 2013). Indeed, while mainstream institutions such as the FAO and the French government have taken up agroecology in their policy discourse, they have also been criticised for reframing agroecology in narrow technocratic terms. The de-politicisation and reshaping of the agroecology discourse risks a conformity to the dominant industrial food and farming system (Levidow et al. 2014; Nyéléni 2015; Holt-Giménez and Altieri 2016; Giraldo and Rosset 2018), in ways that have already played out within the contested framings and definitions of sustainable agriculture, organic agriculture and sustainable development (Guthman 2018; Lélé 1991).

In contrast, food sovereignty is based on the theory that transformative change directly depends on learning and 'development' approaches in which agroecological food producers and citizens are the main protagonists. Farmer organisations and social movements are thus developing and arguing for agroecological 'amplification' processes that are controlled from the bottom up through farmer-driven innovation systems, seed networks, and markets (Nyeleni 2015). Rooted in a food sovereignty perspective, food producers and citizens are claiming their right to define agroecological development (Rosset and Altieri 2017). That said, as a complex, place-specific approach, agroecology presents many challenges to those seeking to understand how it can be up-scaled, extended, or amplified across many locations (Warner 2007).

Agroecology, knowledge and learning: social movement building

Agroecology is knowledge-intensive rather than resource-intensive and thus bolstering knowledge has been claimed as a critical element of any strategy to amplify agroecology. However, formal learning institutions and agricultural extensions generally reflect and institutionalise modernising and elite knowledges that devalue peasant and farmer knowledge (Coolsaet 2016; Meek 2015). Positioned as an alternative paradigm, agroecology as a component of food sovereignty implies an approach to learning that is also transformative in politics and practice.

There is a body of literature on learning that examines processes of learning that allow for personal transformation. Mezirow (1997) defined transformative learning focusing on experiences that have a significant impact on the learner, leading to a paradigm shift that shapes the learner and affects subsequent experiences. While this body of thinking has useful elements in terms of imagining the micro-processes of teaching and learning, it tends to focus on individual learning and change. Whilst individual change is important, agroecology and food sovereignty are being construed as collective political projects that require imagining learning as a strategy of social movement mobilization.

Indeed, social movements advance their own programs of learning and knowledge production that resist the cultural and technical hegemony imposed through agricultural education and other mainstream knowledge institutions (Holt-Giménez 2006; McCune and Sánchez 2019; Freire 1972). Food sovereignty calls for the democratisation of knowledge and learning (Pimbert 2018), and a rejection of the "elite control of education, by also carrying out an agrarian reform of knowledge and culture" (MST 2005, 31 in Barbosa 2016). Knowledge that is collectively developed and held in the autonomous spaces of farmer and community networks is vital for achieving sustainable and just food systems. Social movements have been long articulating this claim including in the recent declaration of the International Forum on Agroecology (Nyéléni 2015):

> Our learning processes are horizontal and peer-to-peer, based on popular education. They take place in our own training centers and territories (farmers teach farmers, fishers teach fishers, etc.), and are also intergenerational, with exchange of knowledge between youth and elders. Agroecology is developed through our own innovation, research, and crop and livestock selection and breeding.

Social movements have rooted much of their learning approaches in a popular education approach (Freire 1972), seeking to raise critical consciousness for transformation, people's empowerment, and the combination of critical theory and political action. Critical pedagogy helps students learn, "to perceive social, political, and economic contradictions, and to take action against the oppressive elements of reality" (Freire 1972). Echoing a long tradition of popular education in Europe (Pelloutier 1901; Gramsci 1971; Ferrer 1996), Scandinavia (Broadbridge et al. 2011), and the Americas (Dewey 1916; Adams and Horton 1975; Freire 1972; McCune and Sánchez 2019), grassroots networks use education as a tool to explore how an individual's personal experiences and learning are linked to larger societal problems and how to transform social and economic injustice in their communities and wider society. Education for critical consciousness usually reflects a deep commitment to radical democracy and human rights (Amsler 2015).

A growing literature, largely from the Latin American context, provides insight into popular education in agrarian movements (Meek 2015; Martínez-Torres and Rosset

2014; Nyéléni 2007, 2015; Wezel et al. 2009; Warner 2007). In the context of food sovereignty, participation social movements is an educational process where it is important to recognise not only the plurality of voices at the grassroots, but as a way to ensure knowledge production is situated there, rather than transmitted from 'above' (Martínez-Torres and Rosset 2014). Focusing on Latin America, Barbosa (2016) highlights the importance of education in the formation of 'the landless identity' as a basis for raising collective consciousness as the basis for social movement mobilisation. There has been minimal attention to how the lessons learned in the historical and cultural context of Latin American resonate with experience in other regions.

European studies are scarce in the literature on grassroots agroecology learning. One exception is Coolsaet (2016) who examines a participatory plant breeding network to demonstrate how agroecology learning can be used as a path towards cognitive justice. Coolsaet joins other commentators in claiming that agroecology necessitates collective learning processes and co-creation of knowledge. However, the study is based on a single case study and also foregoes the development of a more expansive framework commensurate with the scale of agroecology as a global science, movement, and practice. Meek and Tarlau (2016) offer a framework consisting of critical pedagogy, food justice, food sovereignty, and agroecology. While this approach is helpful in developing the linkages between these approaches, our paper focuses on how learning is unfolding on the ground, in the stragegies and processes of social movement building.

This literature provides strong coverage of the Latin American perspective and offers a compelling case for transformative agroecology learning; what it lacks, however, is a coherent framework, one which would be useful to apply at a frontier of agroecology learning, such as Europe or North America. Moreover, while a valorization of local and peasant knowledge and grassroots processes is common in the literature, shifting the locus of education and development to the grassroots does not automatically equate to processes of social transformation. The strategies and processes of transformative empowerment approaches—such as learning and education—are complex. Whilst there is a general sense that popular agroecological education is an important strategy for social movements, there is a need to critically interrogate the mechanisms and theories of change that link popular education to movement building.

The purpose of this paper is to build on the above literature to develop the key principles to guide the implementation of a transformative agroecological learning approach in the hitherto under-examined European context. Our analysis characterizes the emergent European agroecology learning networks and their basis in a critical and strategic knowledge strategy, one based on grassroots knowledge production for the explicit purpose of system transformation.

Methodology

In June 2014, the European Coordination of Via Campesina (ECVC) initiated the formation of a network of agroecology learning in Europe as a strategy to amplify agroecological transformation and to advance food sovereignty (ECVC, n.d.). ECVC is a European network made up of 28 national and regional organisations working to change agriculture policies, defend food sovereignty and peasant rights and create solidarity and exchange between members. EAKEN is an articulation of a wider effort coordinated by La Via Campesina to support the development of Agroecology Learning and Agroecology Schools in the major regions of the world. This article is based on the first stage of the development of this European network which focused on mapping out the dynamics, processes and state of agroecology learning amongst grassroots organisations in Europe.

Interviews

Following an extensive search for participants among ECVC's national 'focal point' contacts, as well as the researchers' own networks, we arranged and conducted 19 semi-structured interviews (Table 1). Specifically, we interviewed leaders from European training initiatives that are directly linked to political movements for food sovereignty and agroecology. These interviews generally lasted between 40 and 90 min, and each participant was asked to describe: (1) the training initiative they are involved in; (2) the key dynamics of agroecology training; (3) the factors supporting or hindering agroecology training; (4) the extent to which government and higher education institutions are involved in agroecology learning; (5) the challenges and opportunities they face in their particular context; and (6) to what extent it would be useful to connect with other agroecology learning initiatives in different parts of Europe.

Analysis and iterative coding process

Interview recordings were transcribed, analysed, and coded in 'Dedoose' qualitative analysis software. Using an iterative approach, our aim was to build up a set of codes to characterise what our interviewees perceived to be the key themes and concepts at play within agroecological learning, especially based on responses to questions 1–4 (above). Each transcript was initially coded using open coding, focusing on reading for immediate meaning in the transcripts. Researchers then reconvened to compare coding structures and a simplified set of 'master codes' were identified.

Table 1 List of interviewees

Interview number	Organisation	Name	Country	Gender
1	Agronauten	Peter Volz	Germany	M
2	Boerengroep	Elske Hageraats	Netherlands	F
3	Butterfly development (Pro-Cserehat Association)	Katalin Rethy	Hungary	F
4	Collectif en Agroécologie Paysanne	Felix Gauthier	France	M
5	Eco Ruralis	Ramona Duminicioiu	Romania	F
6	EHNE-Bizkaia	Ana Gonzalez	Basque Country	F
7	Ileia	Diana Quiroz	Netherlands	F
8	Instituto de Sociología y Estudios Campesinos—ISEC (Universidad de Córdoba)	Mamen Cuellar Padilla	Spain	F
9	L'Atelier Paysan	Julien Reynier	France	M
10	La ligne d'horizon	Sylvia Perez Vitoria	France	F
11	Land Base	Dan Powell	England	M
12	Mouvement Action Paysanne (MAP)	Catherine Tellier	Belgium	F
13	Nordbruk	Joel Holmdahl	Sweden	M
14	Norske bonde-og Smabrukarlag (NBS)	Marielle de Roos	Norway	F
15	The Landworker's Alliance	Lucy Otto	England	F
16	Toekomstboeren	Jildou Friso	Netherlands	F
17	Torth y Tir	Rupert Dunn	Wales	M
18	URGENCI	Gaelle Bigler	Various	F
19	URGENCI	Judith Hitchman	Various	F

At this point, researchers prepared a 'research brief' detailing basic findings structured around these 'master codes', a later version of which can be found on the ECVC website (Anderson et al. 2017a, b). This research brief was taken to the Nyéléni Europe Forum for Food Sovereignty, in Cluj-Napoca, Romania in October 2016 and shared widely, including with those who had already given interviews. Further interviews were conducted at this stage. This began a period of iterative development of our coding structure through participatory workshops, in-person conversations and additional interviews. Over the next 9 months, three further iterations were completed during which we continued to refine our coding structure by identifing redundant aspects as well as those in need of development, before finalising our coding structure. These iterative cycles are detailed in Fig. 1.

Network development

Our participation and contribution to the development of EAKEN have played a key role in our iterative analysis, offering opportunities for further listening and triangulation, and to reflect on the gradual building up of our analysis. Two participatory workshops at the Nyéléni Forum—involving approximately 150 people, including many of our interviewees—were used to recruit participants for the next meeting, a 3 day workshop in the UK. This workshop focused on debating the purpose and functioning of the emerging EAKEN network, and articulating a statement on Agroecology Knowledge Exchange (ECVC 2017). A further two meetings at strategic movement gatherings were used both as an opportunity to maintain momentum within the network but also as to continue the development of our coding structure into something of use to movement practitioners (Fig. 1). In the following sections we present our findings through this framework, in what we have labelled 'four pillars' of a transformative agroecology learning approach. These four pillars are as follows:

1. Diálogo de saberes (wisdom dialogues)
2. Horizontal learning
3. Combining the practical and the political
4. Building multi-scale social movement networks

Agroecology learning initiatives in Europe: Findings

The grassroots organisations represented in our interviews were involved in a wide range of education approaches, including internships, apprenticeships, workshops,

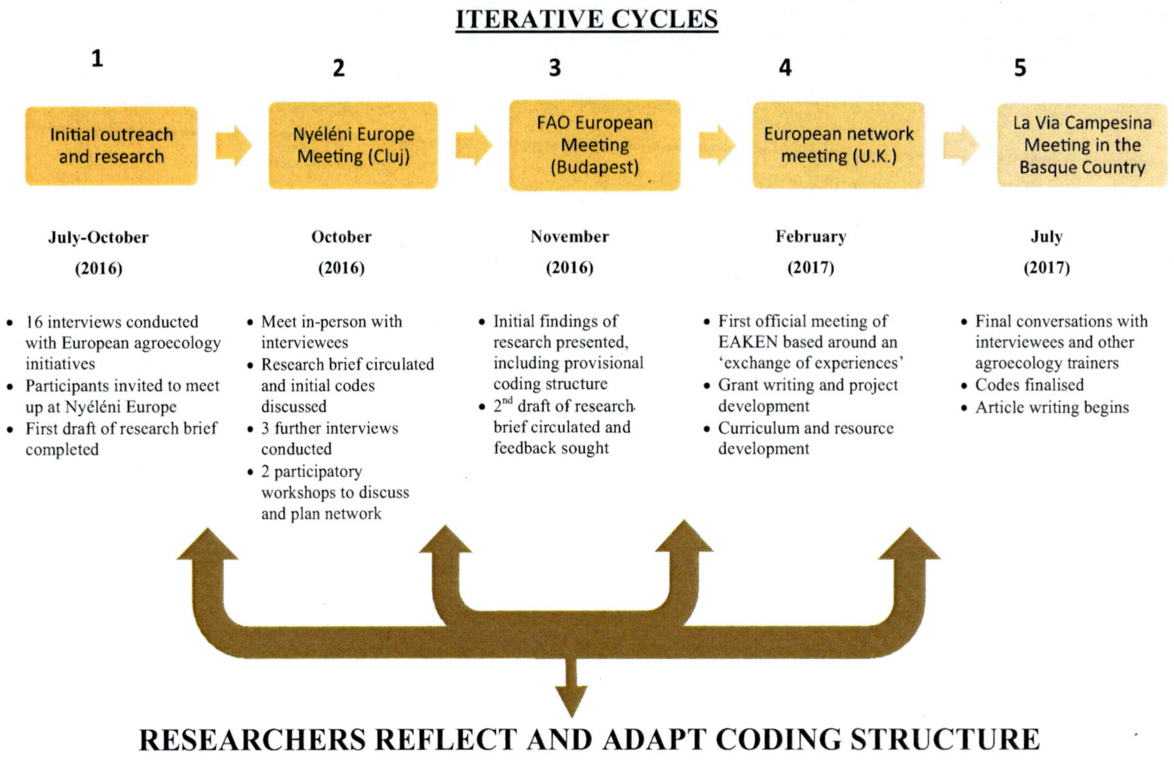

Fig. 1 Iterative stages of research design

skillshares, international gatherings and also joint programs with universities. All of these agroecological initiatives focused on aspects of practical learning, including appropriate technologies, seed saving, increasing soil fertility, animal rearing, marketing, and agroecological farm design. Thus, 'practical' learning about how to effectively implement on-farm agroecological practices were viewed as important to improve agroecological farming systems, enhance the viability of farm businesses and to maximize farm autonomy from external input suppliers.

However, while the practical was central in agroecology learning, it was also clear from our interviews and workshops that agroecological skills and practices only represent one facet of a transformative agroecological learning approach. In the next section, we turn to the additional strategies and characteristics that build on the practical and form the basis of what we call a *Transformative Agroecology Learning*. Our analysis focused on drawing out the approaches, characteristics and methodologies that were most important in linking agroecology with the political project of food sovereignty. This analysis gave rise to four pillars of said approach, and in the following sections, we discuss the general principles of each pillar, illustrating them using examples shared in our interviews and in our action research.

Diálogo de saberes (wisdom dialogues)

Among the representatives of the network that made up our study, a significant emphasis was placed on bringing together actors from different backgrounds. The growing success of rural social movements, and particularly La Via Campesina, has been in part attributed to the ability to foster a *diálogo de saberes* (translated as 'wisdom dialogues' or roughly the equivalent of dialogue between ways of knowing) amongst different rural peoples. While these situated knowledges have emerged through their specific experiences with neoliberal capitalist development, they nonetheless share a broad political interest. Drawing from Boaventura de Sousa Santos (2009, 2010), Martínez-Torres and Rosset (2014) argue that political struggles for food sovereignty and agroecology are based on 'absences' or subaltern knowledge systems that are marginalised by the monopoly of western, corporate and scientific knowledges. In struggles for social change, there are many equally valid ways of knowing the world and transformative learning provides a mechanism for these perspectives to come into dialogue, without one approach dominating another (Martínez-Torres and Rosset 2014). Bringing these together, social movements can produce important 'emergences' at different scales, which form the basis for solidarity, mutual understanding, collective learning and joint action.

In our European research, participants emphasized the importance of these dialogues across three dimensions: Amongst food producers with different positionnings and perspectives; Between food producers and other actors in the food system (especially food consumers); Between food producers and formal education and research institutions.

Amongst Food Producers

Because agroecology is based on a set of underlying ethics and principles which are flexible and not beholden to firm rules or certification, it lends itself to bringing farmers with diverse approaches and worldviews together in a *diálogo de saberes*. As Rupert Dunn, [a peasant farmer and baker from Wales], commented,

> People often don't want to join the club [i.e. any one static and closed system of production] because they feel that they've got to change something about themselves in order to do it. What I like about agroecology is that we don't have to do that [...] I think it incorporates diversity

The agroecology learning initiatives thus brought together different traditions of sustainable agriculture—including permaculture, biodynamics, and organic agriculture. In this way, agroecology functions as a 'boundary object' (i.e. information of different types used in different ways by different communities) which reflects commonly held values of autonomy, food sovereignty, localising food systems and collective knowledges (Anderson et al. 2014).

Many learning initiatives also stressed the importance of bringing together more experienced farmers with new entrants. In the Toekomstboeren program in the Netherlands, for example, an emphasis was put on bringing together an intergenerational network of agroecological producers, which often led to new opportunities for young people to enter farming both in terms of knowledge gained and gaining access to land and resources.

Because of the place-specificity of agroecology, participants argued that inter-place *diálogo de saberes* between people from different biophysical, cultural and political contexts can provide important opportunities for learning and innovation. For example, the importance of facilitating mobility within territories and between them (e.g. in inter-European and international exchanges) was repeatedly mentioned. As Dan Powell, [founder of the Landbase education project in England], stated:

> You just can't learn everything at one place no matter how diverse it is. So to go to different places is really important...in the same way we were discussing about different countries just comparing between different farms. In the same region you could see people doing different things because of their skill based on their position and situation.

Amongst food producers and other actors in the food system

Although much emphasis has been put on farmer knowledge—exchanged especially through farmer-to-farmer learning—examples and debates in EAKEN also suggest that there is a need to emphasize a *diálogo de saberes* between farmers and other actors in the food system, especially consumers. For example, URGENCI, the international network for Community Supported Agriculture (CSA), emphasised the need to consider consumers as key subjects in agroecology and agroecology learning. In one example of their agroecological education work, Ugenci brought farmers and consumers together using theatre as a method to voice concerns, frustrations and dynamics of CSAs. Judith Hitchman, [President of URGENCI] described the potential of extending agroecology learning to work across the farmer-consumer divide, focusing particularly on political education, as a part of a process of building a wider social movement:

> The 2 million people [producers and consumers] in URGENCI all have passive knowledge. The average producer feeds 40 families so you can work out from 2 million people how many producers that is. But the issue is that a great many people have non-political motivation – they find it's very fresh, very high quality it's organic, irrespective of whether it's certified or not. The issue is to bring the political knowledge to grassroots to both producers and consumers – as I said a moment ago they have passive knowledge... What concerns me is how to politicise the cross-cutting grassroots level...

For Judith, the participation of consumers in the political dynamics of agroecology was a rich but largely untapped potential for advancing food system transformation.

Another important *diálogo de saberes* in the network was between farmers and people whose knowledge is being applied in the development of appropriate farm technologies, farm inputs, and market outlets. For example, two of the learning initiatives represented in our initial work reflected a focus on community-developed appropriate technology for production and processing on small scale agroecological farms. Participants discussed how farmer autonomy in agroecology may in many cases be best acheived through interdependence with other actors who share a commitment to food sovereignty. Together they can co-produce innovations that reduce dependency on commercial inputs such as seeds, nutrients or farm machinery. The French cooperative, L'Atelier Paysan, and the UK-based Farm Hack focus on creating on-line and in-person open source learning and innovation communities between small-scale farmers, employees, engineers, software developers and agricultural development organisations. Based on this work, L'Atelier Paysan has begun to articulate and promote the concept of technology sovereignty for farm autonomy, which provides a political framework for the learning (L'Atelier Paysan 2016). This issue is discussed in greater detail in the section on connecting political and practical learning, below.

Between farmers and mainstream research and education institutions

Another important *diálogo de saberes* in agroecology learning can be seen amongst farmers and instructors in the formal education system. Many interviewees attributed some importance to the role of universities and formal education in agroecology, especially given the resources and reach of well-established education institutions. Yet, most had mixed experiences and perspectives reflecting the limiting nature of the institutional and cultural context of formal education in Europe. Participants described, for example, how mainstream institutions were unwilling or incapable of providing training that reflects and includes both the political rootedness and the practitioner-led learning considered essential by social movements.

While some recognised that there were courses on offer within some of these institutions that could be of technical use to agroecological farmer, these were considered generally disconnected from the political and social aspects of the agroecological movement, which rarely took a whole agri-food system approach. For example, Joel Holmdahl, [agroecology educator from Norway] suggested that "We have an increasing number of [...] university courses calling themselves agroecology actually lacking the social and political aspects of agriculture".

A few interviewees mentioned that agroecology courses in universities are primarily aimed at a graduate level, and are often targeted towards agroecological professionals, or experts, rather than farmers. Elske Hageraats [a peasant farmer from the Netherlands], discussing a course on agroecology at Wageningen University, expressed her frustration with the separation of the science of agroecology from the lived experiences of farmers and social movements.

> During the agroecology course, the professors were only looking at the science and I asked: "where is the movement?" and they responded: "yeah well we don't really discuss that". I said: "you should also give students the chance to see first-hand, to visit farms and touch the soil". Someone then told me: "no, here in the university the students are trained to become researchers. If they want to become farmers, they can do that somewhere else"... It seems like most teachers don't see the point of all these connections, or maybe the university doesn't give them the opportunity to teach this. They just want a scientific output. They don't understand the importance of the other two aspects. Students hadn't heard about La Via Campesina, for example, which should definitely be part of it.

Where universities do engage with agroecology, however, interviewees were often critical of the minimal recognition given to farmer knowledge and of the perspectives of social movements, which tended to be viewed as mundane or biased. As Ana Gonzalez [a program coordinator from the Basque Country], from her organisation EHNE-Bizkaia, explained,

> There is a tendency always to say social movements are so influenced by ideology that we normally accused them of being demagogic, but actually what we are trying to say that there is no knowledge in the world that is neutral or objective, and even the academy is based on certain values and principles that were decided by the elites who have decided that this is called academic knowledge.

Despite all these criticisms, however, some farmers' organisations in the network still consider that working with academic institutions can produce mutually beneficial outcomes. Ana went on to describe a healthy collaboration with the University of the Basque Country to develop a Masters program called "from the farm to the world." While it can be difficult to align the priorities and ways of knowing of universities and social movements, Ana viewed this as a process of negotiation leading to interesting outcomes based on a mutual commitment to self-transformation:

> This is about how we are going to create a relationship. When we've finished an educational project [...] neither the social movements nor the universities are the same, we have influenced each other

As Ana argues, however, given the current role of higher education in the accumulation of power in a monopoly of elite knowledge, the potential for social movement engagement with formal institutions must especially involve a transformation of the university itself.

> This is about making the university itself more peasant, more rural...we witnessed that most knowledge and theoretical approaches taught in universities are either directly influenced or lobbied by companies. The content they teach does not question at all the principles on which our economy is based or does not have anything to do with the problems of grassroot communities.

As Ana implies, simply bringing people into dialogue may not be enough, especially if how the dialogue unfolds results in no real questioning of the 'principles on which our economy is based'; some other mechanism or set of principles is needed to ensure one group doesn't simply dominate, and that 'on the ground' realities are adequately represented.

Horizontal learning

The second pillar of a *Transformative Agroecology Learning* approach is Horizontal Learning. or 'Horizontalism' is a central concept within popular education, and involves 'democratic communication on the same level and [... the intention to move towards] non-hierarchical and anti-authoritarian creation rather than reaction' (Sitrin 2006). In this sense, horizontalism can be contrasted to a 'banking' style of education where a teacher deposits expert knowledge into deficient students (Freire 1972). In accordance with this approach, many participants described their intention to recognise the ability of all people to think critically, act strategically, and contribute knowledge. As Jildou Friso [a small scale farmer from the Netherlands] summarised it, *'you shouldn't just be telling people things but leading by example'*. Thus, transformative agroecology learning can be seen to position learners not as the object of teaching, but rather the subjects of their own process of learning, discovery and agency, as well as participants in the joint production of *collective* knowledge throughout their horizontal networks.

Three main aspects of horizontal learning which emerged as important for a transformative learning approach are: (i) strengthening learning experiences; (ii) building confidence and capacity; and (iii) challenging hierarchy through a prefigurative politics.

Strengthening learning experiences

Horizontal learning experiences were promoted by interviewees as producing more effective and durable learning outcomes than top-down knowledge transfer approaches to extension. These approaches often involve bringing the experiences of all learners into dialogue, avoiding single-direction forms of learning. Ashley Trill [a small scale farmer from the U.K.] described a similar dynamic in his experience of farmer-to-farmer learning at a Farm Hack event in the UK:

> I've already learned basically more in the last two hours than I've learned in the last year of being alone weeding or whatever. Everyone here has something to say or a little gem of advice. Tiny little things that are so handy to know.

As with Ashley's sense that all participants had 'a little gem' to contribute, interviewees drew specific attention to the ways in which horizontal approaches fostered trust, genuine engagement and sharing between participants (often farmer-to-farmer) in learner-controlled processes. Ramona Duminicioiu [an activist and organizer from Romania], also stressed the efforts made by her organisation, Eco Ruralis, to take a non-hierarchical approach,

> ...to avoid making peasants feel like they are taught. They have been told for centuries that they are a burden that they are stupid, that they need to be told by the suits from the cities. We are trying to put peasants in situations where they are directing learning experiences.

Similarly, Rupert Dunn [a peasant farmer and baker from Wales] highlighted the importance of farmer-to-farmer dynamics, describing the, "moral support from someone who is a farmer saying, okay try these things out. I've been through that". In addition to the peership dynamics of horizontal learning environments, Rupert also remarked on the simple logic of these types of exchanges:

> it's the only way to have appropriate and tangible accurate knowledges useful for farmers and farms. All farmers test a lot of techniques and systems on their farm and it's really important to be able to scale up all those initiatives and how to scale up is only by bringing people together to exchange it.

A horizontal approach recognises that farmers are more likely to believe and emulate a fellow farmer, than they are to learn production techniques from urban professionals. Due to its peer-to-peer format, the majority of our interviewees echoed Rupert's endorsement of a horizontal agroecology learning approach as a way to validate perspectives routinely sidelined in mainstream learning approaches.

Building confidence and capacity

Horizontal learning environments also have the advantage of fostering confidence in participants, emphasising that everyone has experience and knowledge to contribute in any learning environment (i.e. as teachers), as discussed by Jildou Friso [a small scale farmer from the Netherlands]:

> Something that's really crucial to stimulate the process is to convince people what they know is relevant and that they hold the knowledge themselves.

Katalin Rethy [a program coordinator from Hungary] elaborated on this dynamic in her description of the importance of 'training the trainers'. Such a method, she explained, is essential for scaling up agroecology:

> the trainers that we started to work with four years ago are now capable of training new ones, so it's really about generating local knowledge. We work in Budapest and some of the villages we work with are 150 km away so we are trying to do this knowledge transfer to have local people present in the communities who are employed through our projects to be able to have the knowledge themselves to spread.

Such confidence and capacity building, when embedded within a strategic program to scale up and scale out a learning program, can create a catalytic effect, where more learners become teachers and the process takes on a self-perpetuating momentum (Fakih et al. 2003; McCune and Sánchez 2019).

As Elske Hageraats [a small scale farmer from the Netherlands], pointed out that top-down training or 'big conferences' where 'everyone just presents and then just goes home' can come with a certain 'coldness'; by contrast, horizontal learning environments were consistently presented in terms of the personal connections and emotional processes they initiated, rather than only the instrumental and technical 'outputs':

> the [agroecology] movement is about the feeling, not just the sharing. It is something more, something very deep and I think that has to be touched upon with cultural things like music, sharing, dreaming.

The deliberate fostering of a learning environment which set the grounds for mutual empowerment was a consistent priority amongst our interviewees, especially when bringing different knowledges together in a *diálogo de saberes*. In doing so, interviewees often put emphasis on relationality in the *process*, rather than learning geared only towards output or product. Across the board these were learning experiences centered on personal connections, not merely professional development. Ana Gonzalez [a program coordinator from the Basque Country], for example, described the importance of fostering an "ethos of mutual care" in intergenerational exchanges:

> So it's also about learning from the technical side but also from the relationship side not to judge people. To approach the older generation with more care and a humble attitude, in the way we should always pay attention to history from their side, and how we can set up a more sensitive dialogue with respect and feeling encouraged to teach the new entrants. And sometimes it's about stressing their role as educators. Because when you re-dignify the role, people feel better.

Challenging hierarchy: a prefigurative politics

Interviewees stressed how horizontal learning enacts democratic processes and relations that are consistent with the politics of food sovereignty itself. As one participant put it, food sovereignty is about, "the rights of communities to decide on the food systems they want to build up and participate in" [Ana Gonzalez, a program coordinator from the Basque Country]. As Ana explained, this was often based on a commitment to challenge hierarchy, wherever it is found:

> [our approach] is all about how to make this knowledge more significant for social movements and... for the social struggles ... to challenge to the idea that academic knowledge is the only knowledge that can be considered objective or of any value.

More than a mere set of techniques, horizontalism reflects a commitment to prefigure the kind of bottom-up social change envisioned by the food sovereignty project (Sitrin 2006). In short, horizontal approaches are rooted in the belief in our collective ability to make history and transform society. As such, horizontal pedagogies seek to dissolve the mental blockages or prejudices that often translate into the disabling top-down practices of agricultural research and extension. Indeed, a key focus of horizontal learning approaches is breaking down deeply embedded mental stereotypes that cast farmers and rural people—and especially women—in subservient and helpless roles, even while their capacity to transcend such obstacles has been repeatedly demonstrated (see: Holt-Giménez 2006; Fakih et al. 2003; Machín Sosa 2013; Meek 2015; Rosset et al. 2011; Pimbert et al. 2017; People's Knowledge Editorial Collective 2017).

The political aspects of agroecology may to some seem somehow 'natural' or automatic, yet simply teaching agroecological practice will not automatically equate to political processes of social transformation. As we shall see in the following section the political and practical elements of training must be deliberately and carefully integrated.

Combining the practical and the political

Connecting learning about the practical aspects of agroecological production with the political project of food sovereignty emerged as a third key priority for EAKEN members, particularly as something which was difficult to do or not already being done well. Ramona Duminicioiu [an activist and organizer from Romania], for example, pointed out that,

> Agroecology training must also include the political...I come from a country where peasant farming is a way of life and it's still very vivid in the rural area. So training on how to do agriculture is not a necessary thing. This is happening naturally in the rural communities. But what is missing, and this is what is happening in agroecology, is a more political training on how to articulate our political demands and how to act on achieving political aims.

This emphasis on the political is reflective of attempts to culture political subjectivities through what Barbosa (2016) calls "educational–pedagogical praxis". Such practice emphasizes the central role of education in transforming peasants into historical-political actors as a precondition for political change. As Ana Gonzalez [a program coordinator from the Basque Country] explained, this arises from a pedagogical approach largely based on horizontalism, but also on the deliberate fostering of a lens in learners that critically views all practical problems as political rather than merely technical:

> We want to promote our students to, starting from this critical spirit, become more active or participatory in any other spaces…how are you actually going to do that? By transforming people.

Through collective political dialogic learning, participants can form what Mansbridge and Morris (2001) call collective oppositional consciousness by critically analysing practical problems in the context of a common experience of subordination and to begin to self-identify as a part of a subordinated group. Linking localised learning activities to global discourses of food sovereignty and agroecology provides a basis for participants in social movements. As Julien Reynier [a cooperative organiser from France] from *L'Atelier Paysan* explained, collective political learning can help to overcome individualising tendencies and thus to promote collective subjectivities as the basis for collective action: 'the big issue is the individualism of many farmers today, even alternative farmers [...] they are not involved in farmer collective action'.

Importantly, the practical element of agroecology was for many a strategic way to draw people into political learning that could attract a wider range of participants who may otherwise not engage in such learning or collective action. Julien described this in the case of *L'Atelier Paysan*, an initiative that focuses on collective learning related to farm technology and tools:

> Our strategy is that machinery is quite appealing to farmers and we don't have to struggle that much to bring people together around this topic...every farmer needs good machinery. And it's really important to increase the viability of your farm...That's why we think being more technical gives you the opportunity to get involved and at the same time we need people who can speak to the heart of the farmers and get them involved in political actions.

Similarly, EHNE-Bizkaia's agroecology training program in the Basque Country often leads with practical skills, moving on when participant confidence has increased to incorporate discussions on social movements and politics. This reflects a key aspect of critical pedagogic practice that starts with the knowledge and practical problems of the learner, but through dialogue and reflection locates this knowledge in the wider political and economic context. In the Romanian context, Ramona [an activist and organizer from Romania] described how the confidence and capacity for political analysis builds over time, increasing the critical consciousness of learners and their capacity to act on this new understanding.

> We discovered that people who in the past were afraid to say anything regarding political issues are now curious and more interested and able to make connections between different discussions and agendas.

Learning on agroecological practice, in essence, can act as an on-ramp to engage with the politics of food sovereignty, which can then be translated into political action, as described by Peter Volz [a researcher from Germany],

> ...the second step is to enable action on a personal level, on a collective level but anyhow in the frame that these people are operating within.

While the belief in integrating the practical and the political was widely held amongst the people we spoke with, interviewees in many cases were less able to articulate precise approaches and methodologies for enabling this integration. Indeed, many of them pointed out that the simple act of gathering farmers together was a key strategy for political learning—"get farmers together and they will invariably turn to politics" [Diana Quiroz, a coordinator from the Netherlands].

While creating convivial space with time for free conversations on the occasions of practical training may lend itself to the emergence of political learning, there are also more explicit and systematic methods to enhance political learning. For instance, EAKEN has created an both face-to-face and online spaces to cooperatively analyse and systematically share pedagogic examples, methods and resources to build capacity in combination of practical and political learning approaches.

Building multi-scale social movement networks

The final pillar of the Transformative Agroecological Learning approach focuses on building social movement networks. The horizontal scaling out of grassroots agroecological learning is largely driven by the processes identified thus far in this paper: critical education, *diálogo de saberes*, horizontal networks for peer-to-peer learning, and an emphasis on collective practical and political knowledge. All these processes depend on coordinated action made possible by local organizations that bring people together for joint activities—from resource management, labour-sharing, marketing, advocacy and other activities that would be too costly, or impossible, if done alone.

To this end, Guy Kastler [a peasant farmer from France], commented at one of the EAKEN meetings that, "there is no agroecology knowledge without strong farmer organisations", echoing the idea that agroecological knowledge is always collective (Nyéléni 2015). It is in organisations and networks that a knowledge commons is built and strengthened, and linked to the material and political project of food sovereignty.

All major success stories in grassroots agroecological education depend on nested local organizations to facilitate and coordinate collective action at different scales (Pimbert 2008, 2018). Decentralized and distributed local organizations play a key role in facilitating knowledge-sharing within transformative agroecological learning networks. Interlinked local organizations and collective structures provide the institutional landscape that enables the coordinated and timely exchange of agroecological knowledge within networks of food producers and consumers. This suggests that local organizations—and the networks they form—need to be strengthened in order to enhance the capacity to 'scale out' the transformative agroecological learning processes described here.

In many cases, participants in the network spoke of education as an intentional strategy to strengthen local organisations, contributing to what has been called a territorialisation of agroecology (McCune and Sánchez 2019; Wezel et al. 2016). As Julien Reynier [a cooperative organiser from France] from *L'Atelier Paysan* put it, "the method is trying to build communities".

Agroecology training was also viewed as important in building national political networks, where, for example, learning initiatives like Land Base and Farm Hack in the UK were linked to, and help to enrol new participants into the Land Workers Alliance (a UK member of La Via Campesina). Ramona [an activist and organizer from Romania] described how their combination of practical and political learning helps to expand enrolment in the Romanian member organisation Eco Ruralis:

> Political issues are always part of these meetings in a transversal way…what is happening with legislation around seeds or GMOs, what can we do in Romania, if we need to go on alert or call our MEPs …and slowly, in a transversal way, it builds a flow of information, and it builds slowly a base.

Network building through transformative agroecology learning was multi-scaled in nature: EAKEN, which aims to support local, regional and national initiatives is connected to a global strategy by La Via Campesina to develop capacity in agroecology learning approaches, methodologies and experiences. By building international networks, new possibilities arise for exchange and the international mobility of pedagogical innovations from one location to another through a *dialogos de saberes*. For example, learning from Latin American social movements was important for EHNE-Bizkaia which now has one of the most extensive agroecology education programs among all the initiatives currently in EAKEN. Ana Gonzalez [a program coordinator from the Basque Country] explained,

> The main strategy for EHNE-Bizkaia, and our experience with La Via Campesina has allowed us to learn from others how to do things or how to set up educational trainings which were creating a different perspective, we have received much more influence from Cuba on how to set up educational programs from the perspective of agroecology

These multi-scaled networks are critical for the sharing of agroecological knowledges and pedagogies across places, building capacity for further learning. As Mamen Cuéllar Padilla [an activist-research from Spain] described, these networks provide an opportunity to support innovation in pedagogic approaches, especially in terms of the transformative aspects of agroecology learning:

> We can exchange proposals, thinking, methodologies, contents, materials, documents, experiences and practice but also it could help to strengthen this vision of knowledge dialogue, horizontal power relations, the role of academia in this knowledge creation and this training action.

Learning programs linking farmers to organisations that are advancing food sovereignty was also viewed as a key element of developing a sense of collective consciousness. Participation in these organisations affirms the identity and the work of peasants and agroecologists, strengthens commitment to advancing agroecology and provides a sense of belonging to a wider social movement. As Jildou Friso [a peasant farmer from the Netherlands] described,

> ...it's such a struggle to go on with agroecology practice. Just the feeling of being a part of a network is very empowering. If you share your knowledge and people from other countries are interested in what you're doing it can be very empowering – it's a small but important point.

Most of the initiatives in this study were connected to La Via Campesina who are articulating strong positions and campaigns for food sovereignty. Agroecological learning can connect learners into these multi-scaled networks and organisations which act as mobilising structures for political work at local, national and international levels. These organisations provide critical mechanisms for exercising dissent and countervailing power, which is a mainstay of social movement practice (Freire 1972). However, these networks were also charachterised by uneven participation in these learning and knowledge exchange opportunities. As Judith Hitchman [President of URGENCI] suggested,

> Even with all the European projects we're doing, to a large extent it's the top 10% of the leadership. What concerns me is how to politicise and get the knowledge and training down to the cross-cutting grassroots level.

While we will address this in greater detail in the following section, many interviewees saw the combining of practical and political learning as a way to address such unevenness; that is, as a means to sustain the link between abstract political work and particularised agroecological practice. Ramona [an activist and organizer from Romania] elaborated,

> We need to keep both connected because people that are getting involved in political actions tend to lose the connections with people that are involved in practical activities.

Discussion

The initiatives and approaches we characterised in this paper as a part of EAKEN are explicitly developing agroecology learning as a method of social transformation. In this context, learning was couched for the specific purpose of enabling transformation and was a strategy for social movement mobilisation. In addition to technical learning, these citizen-led initiatives base their work on four additional pillars: Horizontalism; Diálogo de saberes; Combining practical and political; and building social movement networks (Fig. 2).

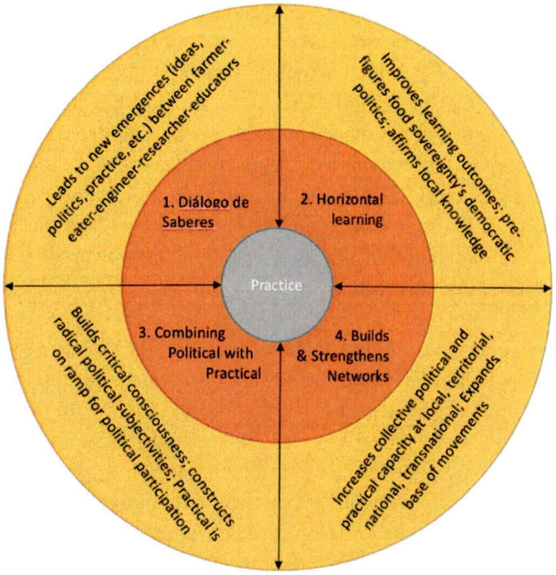

Fig. 2 Transformative agroecology framework involves a pedagogical approach that always has practice as a central component, but also relies on four pillars (orange segments) to provide the 'connective tissue' to the political project of food sovereignty (yellow circle). (Color figure online)

While we have presented the four pillars of agroecology learning as distinct, our findings also provide insight into the interdependencies and tensions that arise within and between them. Indeed, these dynamics are a crucial part of advancing current thinking on transformative agroecology learning. For example, the expanded vision of a multi-actor *diálogo de saberes* is a frontier for the agroecology movement in Europe. That is, opening up opportunities for learning between farmers and other constituents such as citizen-food-consumers, researchers and educators comes with risks, as the positionality of different actors implies intersecting and uneven relations of power. Indeed, our results reiterated how farmer knowledge is often structurally and culturally subordinated by urban and professional knowledges and interests. Thus, these dialogues (pillar 1) must be developed with caution and critical reflexivity. To this end it is essential to establish mutually agreed upon protocols and pedagogies based on horizontalism (pillar 2) to preempt the emergence of exploitative hierarchies. For example, in the early development of EAKEN, there have been careful steps taken to ensure that farmers have safe spaces to articulate their own priorities and needs, to focus on dialogue amongst different food producers, and to strategically engage in these wider dialogues as the opportunities to do so are developed.

Yet, the challenge remains that while there is a recognition of the need for these wider cross-constituent dialogues and partnerships, action too easily falls back into old silos

(Anderson et al. 2017a, b). Indeed, our framework encourages work that is embedded in social movement organisations (pillar 4) which are often based on single constituencies (e.g. peasants). This suggests that the ability to foster cross-constituency dialogues may require network and organisational innovations to create new spaces for this dialogue to occur, based on horizontal relations. This is fundamentally important to building solidarity and conditions for allyship between differently positioned actors in the food sovereignty movement.

One emergent, yet tension-ridden *diálogo de saberes* was between food producers and other actors of the food system which reflects the increasing, but unmet, call to move from a farm-level focus to a whole food system approach to agroecology (Wezel et al. 2009, 2016). This systems approach recognizes that transforming the food system will be impossible without the practical and political participation of non-producers (González de Molina 2013) (i.e. consumers). In contrast to research on *diálogo de saberes* in the global south where the focus was on rural-constituents (Martínez-Torres and Rosset 2014), this wider dialogue may be especially salient in a highly urbanised Europe where significant challenges remain in making food sovereignty relevant to the lives of urban non-producing people. Based on a weaving of practical and political learning between farmers, food workers and consumers, a transformative agroecology learning may address some of the critiques of alternative food networks (which link producers and consumers) and their tendency to depoliticise consumers (e.g. Johnston et al. 2009). Conceptualising these interactions as potential moments of critical and political learning remains an important opportunity to mobilize people who buy food from agroecology farmers as more-than-consumers, but also as "historical–political actors" (Barbosa 2016) in struggles for food sovereignty.

An example of a positive synergy between the pillars, is the ability of a horizontal approach (pillar 2) to build capacity and confidence amongst learners, who can become active participants, organisers and leaders in social movement organisations (pillar 4). A top-down approach to learning can foster passivity and reinforce hierarchies between expert teacher-leaders and lay learner-followers, thus undermining the potential to expand participation in developing social movement organisations. By providing opportunities for learners to begin to provide active input and leadership in organizations (rather than only provide instrumental input or token involvement) through applying their political learning in organising campaigns or becoming facilitator-teachers in learning initiatives, these networks can be expanded and strengthened, as demonstrated in the case of the Cuban ANAP and Brazilian MST movements (McCune and Sánchez 2019; Rosset et al. 2011).

Across the network, there were clear differences amongst the organisations in terms of whether transformative agroecology learning was used as an explicit organisational strategy. For some, agroecology learning was a core and integrated element of organisational strategies. For others, it was minor and disconnected from other activities, such as advocacy. If transformative agroecology learning is genuinely to provide the 'connective tissue' between food sovereignty and agroecology, there is a need to: (a) strategically plan to incorporate it as a part of an integrated organisational development strategy; (b) contribute resources to the systematisation and improvement of transformative agroecological learning praxis. In regards to the latter, the four pillars were unevenly developed across EAKEN member organisations, and in many cases the precise methods and approaches to implementing the four pillars were difficult for many of the interviewees to articulate.

While a number of understandable factors (such as funding constraints, recent establishment, and minimal integration into international networks) contributed to this unevenness, there is a danger that, without systematisation and critical reflection, such organisations might easily fall back into dominant modes of learning (e.g. technical, top-down) and de-prioritize bottom-up transformative learning as an indispensable mode of social movement building. Compared to more conventional agroecology learning, these pillars require the continual systematic reappraisal, redesign and re-invention in response to the evolving dynamics of the organisation (e.g. as participant composition evolves) and changes in the wider context (e.g. as the food crisis develops).

The systematisation of transformative agroecological learning approaches is one of the key purposes of EAKEN, as well as the global articulation coordinated by La Via Campesina. These connections, between the European and Global network can provide opportunities for organisation-to-organisation learning, especially from regions and organisations that have advanced further. For example, in the MST, an education sector and the [political] formation sector create organisational structures to intentionally develop the linkages between education and movement building (McCune and Sánchez 2019). Networks like EAKEN, we argue, are key to facilitating inter-organisational learning and helping to more systematically incorporate transformative learning into social movement strategies.

While agroecology learning is relatively embryonic across Europe, its potential as a tool to build up social movement networks has been exemplified in the Cuban *Campesino-to-Campesino* movement, where, through social processes of learning, the program built up a base made up of 50% of all the peasant families in Cuba[1] (Rosset et al. 2011).

[1] A multimedia resource for learning about the Cuban experience can be found here: http://agroecologia.espora.org/.

Of course, there are many differences between the European context and the Cuban or other Latin American countries where much of the literature on agroecology learning has been developed. First, there is generally a larger peasant contingent in the Latin American context, with denser networks. Next, there is a much stronger tradition of popular education, whereas in Europe these traditions have been weakened through the neoliberal period. Finally, whereas in many Latin American contexts, peasant social movements have long advanced agroecology within an anti-capitalist and decolonizing framework, peasant movements in Europe have only recently began to embrace a transformative agroecology for food sovereignty (Nyéléni 2017; Pimbert 2015). There are also a wide range of powerful governmental, scientists and non-governmental institutions that are competing to define a European articulation of agroecology—including competing notions of the practices and politics of learning and training. In this context, the meaning and the doing of agroecology is far more ambiguous and up for grabs, making the transformative agroecology learning more urgent, yet also more challenging.

Conclusion

Our analysis contributes to the theory of learning for transformative agroecology based on four key processes: horizontalism; *diálogo de saberes* (wisdom dialogues); combining practical and political knowledge; building social movement networks. While theories of transformative learning often focus on individual learning processes, our approach was to conceptualise learning as an integral part of the process of social movement building. This resonates strongly with other contributions in this special issue where learning focuses on territorial integration (McCune and Sánchez 2019) or a process of local dynamization (López-García et al. 2019). In this context, the four key processes described here could provide a tool to for social movements to strategically and reflexively systematize and strengthen a transformative learning approach. While this was applicable in the context of agroecology for food sovereignty, this framework may also be relevant for other movements.

The research rasises further questions about the nature of transformative learning. For example, in terms of the systematisation of current practice, what are the organisational, methodological and institutional innovations that can support bottom up learning and critical education in a durable way? How can state funding shift from the current emphasis on top-down technological fixes, towards appropriately supporting the decentralised social learning processes being advanced by social movements? Similarly, how can farmers and other actors meaningfully participate in transformative agroecological learning on a long-term basis given the precarity of their own material circumstances (Calvário 2017; Pimbert 2011). Although we have focused on collective learning processes as a social movement strategy, further work is needed to understand how this social-political pedagogy is connected to individual transformative learning processes (e.g. Mezirow 1997).

Next, the actual processes and methods for enabling *diálogo de saberes* and horizontalism in agroecology learning requires further close scrutiny, especially the role of educators and facilitators, to better understand what attitudes, behaviors and skills are required for a transformative learning approach. More work is needed to understand how learning approaches should be, like agroecology practices themselves, adapted to contingencies of place including the biophysical, organisational, historical, cultural and institutional differences between places. While it is recognised that the potential of a transformative agroecology learning will only be realised to the extent that it contributes to territorial processes of agroecological dynamisation (McCune and Sánchez 2019; López-García et al. 2019), there is little known about how to adapt to differential territorial contexts. Finally, many questions remain to better understand the specific processes within each of the domains and further in-depth research is needed to examine, for example, how to most effectively engage a wide diversity of actors—often with uneven relations of power—in learning for transformation.

The systematisation of a transformative agroecology learning is a critical step for regions, movements and groups seeking to advance the project of food sovereignty. This article offers a contribution to this process of systematisation and highlights a range of evolving processes from the European context. Through collective learning that links agroecology practices with the political project of food sovereignty, social movements are advancing a broader transformation towards a more just and sustainable food system.

References

Adams, F., and M. Horton. 1975. *Unearthing seeds of fire: the idea of Highlander*. Winston Salem: John F. Blair.

Aerni, P. 2011. Food sovereignty and its discontents. *ATDF Journal* 8 (1–2): 23–40.

AgriCultures Network. 2013. Brazil launches agroecology plan. AgricCultures Network. http://www.agriculturesnetwork.org/news/brazil-agroecology-plan. Accessed 27 July 2016.

Alonso-Fradejas, A., S. M. Borras Jr, T. Holmes, E. Holt-Giménez, and M. J. Robbins. 2015. Food sovereignty: convergence and contradictions, conditions and challenges. *Third World Quarterly* 36 (3): 431–448.

Altieri, M. 2009. Agroecology, small farms, and food sovereignty. *Monthly Review* 61: 102–113

Altieri, M. A., and V. M. Toledo. 2011. The agroecological revolution in Latin America: rescuing nature, ensuring food sovereignty and empowering peasants. *Journal of Peasant Studies* 38 (3): 587–612.

Amsler, S. S. 2015. *The education of radical democracy*. New York: Routledge.

Anderson, C. R., W. McDonald, J. Gardiner, and S. McLachlan. 2014. Navigating the fault lines in civic food networks. *Journal of Agriculture, Food Systems and Community Development*, 4 (3): 79–99.

Anderson, C. R., Silivay J. and K. Lobe. 2017a. Community organisations for food systems change: reflecting on food movement dynamics in Manitoba. In *Everyday experts: how people's knowledge can transform the food system*, ed. Peoples Knowledge Collective. Coventry: Coventry University.

Anderson, C., C. Maughan, and M. Vizy. 2017b. Developing a European agroecology learning and training network: research brief. Centre for Agroecology, Water and Resilience and the European Coordination of Via Campesina. https://www.eurovia.org/eaken/. Accessed 18 September 2018.

Barbosa, L. P. 2016. Educação do Campo [education for and by the countryside] as a political project in the context of the struggle for land in Brazil. *The Journal of Peasant Studies* 44 (1): 118–143.

Bernstein, H. 2014. Food sovereignty via the "peasant way": a sceptical view. *Journal of Peasant Studies* 41 (6): 1031–1063.

Broadbridge, E., U. Jonas, and C. Warren. 2011. *The school for life: NFS Grundtvig on education for the people*, vol. 1. Aarhus: Aarhus Universitetsforlag.

Calvário, R. 2017. Food sovereignty and new peasantries: on re-peasantization and counter-hegemonic contestations in the Basque territory. *The Journal of Peasant Studies* 44 (2): 402–420.

Coolsaet, B. 2016. Towards an agroecology of knowledges: recognition, cognitive justice and farmers' autonomy in France. *Journal of Rural Studies* 47 (A): 165–171.

De Sousa Santos, B. 2009. *Una epistemología del sur: la reinvención del conocimiento y la emancipación social*. Siglo XXI.

De Sousa Santos, B. 2010. *Descolonizar el saber, reinventar el poder*. Ediciones Trilce.

Desmarais, A. A. 2007. *La Vía Campesina: globalization and the power of peasants*. Halifax: Fernwood.

Dewey, J. 1916. *Education and democracy: an introduction to the philosophy of education*.

ECVC. European Coordination of Via Campesina. n.d. European Agroecology Knowledge Exchange Network. https://www.eurovia.org/eaken/. Accessed 18 September 2018.

ECVC. 2017. Monkton wyld statement on agrecology knowledge exchange. ECVC. Retrieved from http://www.eurovia.org/wp-content/uploads/2017/02/2017-02-Monkton-Wylde-Statement-on-Agroecology-Knowledge-Exchange.pdf.

Edelman, M. 2014. Food sovereignty: forgotten genealogies and future regulatory challenges. *The Journal of Peasant Studies* 41 (6): 959–978.

Fakih, M., T. Rahardjo, and M. P. Pimbert. 2003. *Community integrated pest management in Indonesia: Institutionalising participation and people centered approaches*. London: International Institute for Environment and Development (IIED), Institute for Development Studies (IDS), and REaD.

FAO. 2018. *Catalysing dialogue and cooperation to scale up agroecology: Outcomes of the FAO Regional Seminars on Agroecology*. Rome: FAO. http://www.fao.org/3/I9035EN/i9035en.pdf. Accessed 19 September 2018.

Ferrer, A. T. 1996. The workers' movement and popular education in contemporary Spain (1868–1939). *Paedagogica Historica* 32 (3): 647–684.

Francis, C., T. A. Breland, E. Østergaard, G. Lieblein, and S. Morse. 2013. Phenomenon-based learning in agroecology: a prerequisite for transdisciplinarity and responsible action. *Journal of Sustainable Agriculture* 37 (1): 60–75.

Freire, P. 1972. *Pedagogy of the oppressed*. New York: Herder.

Giraldo, O. F., and P. M. Rosset. 2018. Agroecology as a territory in dispute: between institutionality and social movements. *The Journal of Peasant Studies* 45 (3): 545–564.

Gliessman, S. R. 2014. *Agroecology: the ecology of sustainable food systems*. Boca Raton: CRC Press.

González de Molina, M. 2013. Agroecology and politics. How to get sustainability? About the necessity for a political agroecology. *Agroecology and Sustainable Food Systems* 37 (1): 45–59.

Gramsci, A. 1971. *Selections from the Prison Notebooks of Antonio Gramsci*. Edited and Translated by Quintin Hoare and Geoffrey Nowell Smith. International Publishers.

Guthman, J. 2018. *Agrarian dreams: the paradox of organic farming in California*. 2nd ed. Oakland: University of California Press.

Holt-Giménez, E. 2006. *Campesino a campesino: voices from Latin America's farmer to farmer movement for sustainable agriculture*. Oakland: Food First.

Holt-Giménez, E., and M. Altieri. 2016. *Agroecology "lite:" cooptation and resistance in the global north*. Food First. https://foodfirst.org/agroecology-lite-cooptation-and-resistance-in-the-global-north/. Accessed 19 September 2018.

IAASTD (International Assessment of Agricultural Knowledge, Science and Technology for Development). 2009. *Synthesis report: a synthesis of the global and sub-global IAASTD reports*. Washington: IAASTD.

Johnston, J., A. Biro, and N. MacKendrick. 2009. Lost in the supermarket: The corporate-organic foodscape and the struggle for food democracy. *Antipode* 41 (3): 509–532.

L'Atelier Paysan. 2016. Plaidoyer: souveraineté technologique des paysans. http://www.latelierpaysan.org/Plaidoyer-souverainete-technologique-des-paysans. Accessed 19 July 2017.

La Via Campesina (LVC). 2013. *From Maputo to Jakarta: 5 years of agroecology in La Vía Campesina*. Jakarta. http://viacampesina.org/downloads/pdf/en/de-maputo-a-yakarta-en-web.pdf. Accessed 1 January 2018.

Lang, T. 2010. Crisis? What crisis? the normality of the current food crisis. *Journal of Agrarian Change* 10 (1): 87–97.

Lélé, S. M. 1991. Sustainable development: a critical review. *World Development* 19 (6): 607–621.

Levidow, L. 2015. European transitions towards a corporate-environmental food regime: agroecological incorporation or contestation? *Journal of Rural Studies* 40: 76–89.

Levidow, L., M. P. Pimbert, and G. Vanloqueren. 2014. Agroecological research: conforming—or transforming the dominant agro-food regime? *Agroecology and Sustainable Food Systems* 38 (10): 1127–1155.

López-García, D., L. Calvet-Mir, M. Di Masso, and J. Espluga. 2019. Multi-actor networks and innovation niches: university training for local agroecological dynamization. *Agriculture and Human Values*. https://doi.org/10.1007/s10460-018-9863-7.

Machín Sosa, B. 2013. *Agroecological revolution: the farmer-to-farmer movement of the ANAP in Cuba [online]*. Havana and Jakarta: ANAP and La Vía Campesina.

Mansbridge, J. J., and A. Morris. 2001. *Oppositional consciousness: the subjective roots of social protest*. Chigago: University of Chicago Press.

Martínez-Torres, M. E., and P. M. Rosset. 2014. Diálogo de saberesin La Vía Campesina: food sovereignty and agroecology. *The Journal of Peasant Studies* 41: 979–997.

McCune, N., and M. Sánchez. 2019. Teaching the territory: agroecological pedagogy and popular movements. *Agriculture and Human Values*. https://doi.org/10.1007/s10460-018-9853-9.

McMichael, P. 2008. Peasants make their own history, but not just as they please. *Journal of Agrarian Change* 8 (2–3): 205–228.

Meek, D. 2015. Learning as territoriality: the political ecology of education in the Brazilian landless workers' movement. *The Journal of Peasant Studies* 42 (6): 1179–1200.

Meek, D., and R. Tarlau. 2016. Critical food systems education (CFSE): educating for food sovereignty. *Agroecology and Sustainable Food Systems* 40 (3): 237–260.

Méndez, V. E., C. M. Bacon, R. Cohen, and S. R. Gliessman. 2015. *Agroecology: transdisciplinary, participatory and action-oriented approach*. Boca Raton: CRC Press.

Mezirow, J. 1997. Transformative learning: theory to practice. *New Directions for Adult and Continuing Education* 74: 5–12.

Mier y Terán Giménez Cacho, M., O. F. Giraldo, M. Aldasoro, H. Morales, B. G. Ferguson, P. Rosset, A. Khadse, and C. Campos. 2018. Bringing agroecology to scale: key drivers and emblematic cases. *Agroecology and Sustainable Food Systems* 42 (6): 637–665.

MST. 2005. Dossiê MST - Escola. Documentos e estudos 1990–2001. In Caderno de Educação, n° 13. 2a Edição. São Paulo: Expressão Popular.

Nyéléni. 2007. Nyéléni 2007—forum for food sovereignty—full report. Nyéléni—forum for food sovereignty. http://Nyeleni.org/DOWNLOADS/Nyelni_EN.pdf. Accessed 18 September 2018.

Nyéléni. 2015. Nyéléni 2015—Declaration of the international forum for agroecology. Nyéléni—forum for food sovereignty. http://www.foodsovereignty.org/forum-agroecology-Nyeleni-2015/. Accessed 18 September 2018.

Nyéléni Europe. 2017. Nyéléni pan-European forum for food sovereigty, Cluj Napoca—Romania—Expo Transilvania, 25–30 October 2016. Cluj Napoca—Romania: Nyéléni Europe. http://Nyelenieurope.net/sites/default/files/2017-04/Nyeleni%20Europe%20Report%202016_web.pdf. Accessed 18 September 2018.

Patel, R. 2009. Food sovereignty. *The Journal of Peasant Studies* 36 (3): 663–706.

Pelloutier, F. 1901. *Histoire des bourses du travail*. London: Routledge.

People's Knowledge Editorial Collective (Eds). 2017. *Everyday experts: how people's knowledge can transform the food system*. Coventry: Coventry University.

Pimbert, M. P. 2008. *Towards food sovereignty: reclaiming autonomous food systems*. London: International Institute for Environment and Development (IIED).

Pimbert, M. P. 2011. *Participatory research and on-farm management of agricultural biodiversity in Europe*. London: International Institute for Environment and Development (IIED).

Pimbert, M. P. 2015. Agroecology as an alternative vision to conventional development and climate-smart agriculture. *Development* 58 (2–3): 286–298.

Pimbert, M. P., P. V. Satheesh, A. Argumedo, and T. M. Farvar. 2017a. Participatory action research transforming local food systems in India, Iran and Peru. In: People's Knowledge Editorial Collective (Eds). Everyday Experts: How people's knowledge can transform the food system. Reclaiming Diversity and Citizenship Series. Coventry: Coventry University. http://www.coventry.ac.uk/everyday-experts.

Pimbert, M. P. 2018. *Food sovereignty, agroecology and biocultural diversity. Constructing and contesting knowledge*. London: Routledge.

Rosset, P. M., and M. A. Altieri. 2017. *Agroecology, science and politics*. Winnipeg: Fernwood.

Rosset, P. M., B. Machín Sosa, A. M. Roque Jaime, and D. R. Ávila Lozano. 2011. The Campesino-to-Campesino agroecology movement of ANAP in Cuba: social process methodology in the construction of sustainable peasant agriculture and food sovereignty. *Journal of Peasant Studies* 38 (1): 161–191.

Sitrin, M. 2006. *Horizontalism: voices of popular power in Argentina*. Oakland, CA: AK Press.

Tittonell, P., L. Klerkx, F. Baudron, G. F. Félix, A. Ruggia, D. van Apeldoorn, S. Dogliotti, P. Mapfumo, and W. A. Rossing. 2016. Ecological intensification: local innovation to address global challenges. In Sustainable Agriculture Reviews (pp. 1–34). Springer.

Warner, K. 2007. *Agroecology in action: extending alternative agriculture through social networks*. Cambridge, Mass: MIT.

Wezel, A., S. Bellon, T. Doré, C. Francis, D. Vallod, and C. David. 2009. Agroecology as a science, a movement and a practice. A review. *Agronomy for Sustainable Development* 29 (4): 503–515.

Wezel, A., H. Brives, M. Casagrande, C. Clément, A. Dufour, and P. Vandenbroucke. 2016. Agroecology territories: places for sustainable agricultural and food systems and biodiversity conservation. *Agroecology and Sustainable Food Systems* 40 (2): 132–144.

Wittman, H. 2011. Food sovereignty: a new rights framework for food and nature? *Environment and Society* 9 (1): 87–105.

Colin R. Anderson is a Senior Research Fellow at the Centre for Agroecology, Water and Resilience, Coventry University. His research focuses on community- and people- led processes of transformation for resilience, social justice and well-being. He works with communities, networks and organizations who are adressing the multitude of social, economic and cultural problems we face today and who are reimaging society and building alternatives. His collaborative research contributes to the development of thinking and action related to food systems, community building, food sovereignty, activism, and knowledge mobilization, and especially focuses on the workings of power, resilience, and regeneration in these areas. Visit https://www.agroecologynow.com and https://www.peoplesknowledge.org to learn more.

Chris Maughan is a British Academy Postdoctoral Fellow based at the Centre for Agroecology, Water and Resilience. His research interests include the role of popular education, grassroots innovation processes, and inclusive governance in the development and realisation of agroecological food systems. He is currently leading on a research project designed to develop the latest theory on innovation processes and draw from bottom-up knowledge exchange processes and community economies theory to propose new ways to scale-up agroecological innovation. This project draws together the analysis of bottom-up processes of horizontal knowledge exchange and interactive innovation in agroecology networks, in order to develop a more holistic theory of innovation that can better harness the economic, social, cultural and political processes needed to develop a just and sustainable food system.

Michel P. Pimbert is Professor of Agroecology and Food Politics at Coventry University and the Director of the Centre for Agroecology, Water and Resilience in the UK. His research interests include agroecology and food sovereignty; the political ecology of biodiversity and natural resource management; participatory action research methodologies; and deliberative democratic processes. He works with networks of small and family farmers, indigenous peoples, and communities to advance transdisciplinary ways of knowing that regenerate local ecologies, economies and cultural diversity.

Agriculture and Human Values (2019) 36:549–566
https://doi.org/10.1007/s10460-018-09906-x

SYMPOSIUM/SPECIAL ISSUE

Farming for change: developing a participatory curriculum on agroecology, nutrition, climate change and social equity in Malawi and Tanzania

Rachel Bezner Kerr[1] · Sera L. Young[2] · Carrie Young[3] · Marianne V. Santoso[4] · Mufunanji Magalasi[5] · Martin Entz[6] · Esther Lupafya[7] · Laifolo Dakishoni[7] · Vicki Morrone[8] · David Wolfe[9] · Sieglinde S. Snapp[10]

Accepted: 18 December 2018 / Published online: 28 January 2019

Abstract
How to engage farmers that have limited formal education is at the foundation of environmentally-sound and equitable agricultural development. Yet there are few examples of curricula that support the co-development of knowledge with farmers. While transdisciplinary and participatory techniques are considered key components of agroecology, *how* to do so is rarely specified and few materials are available, especially those relevant to smallholder farmers with limited formal education in Sub-Saharan Africa. The few training materials that exist provide appropriate methods, such as compost making, but do not explain relationships and synergies between nutrition, social inequalities, climate change and agroecology. Some food sovereignty and agroecology courses aim at popular political education for those with more formal education. Here we describe the process of development of an innovative curriculum, which integrates agroecology, nutrition, climate change, gender and other dimensions of social equity across 2 weeks of training explicitly for smallholders in southern Africa with limited formal education. The curriculum is highly participatory; we use concepts in popular education, transformative and experiential-based learning, and theatre. It is also integrative; we link agroecology with climate change, human and soil nutrition, gender, and related components of social equity. Developed in partnership with Malawian farmers, community development experts and academics from five countries, the curriculum was piloted with 520 smallholder farming households in Malawi and Tanzania, and evaluated using qualitative techniques. Clashes of language, cultural norms, and terminology were as great of a challenge as agreeing on and conveying technical information, to weave into a coherent whole. However, farmers who participated in the curriculum training demonstrated high interest, comprehension of material and interest in immediate application to their lives.

Keywords Critical food systems education · Agroecology · Transdisciplinary · Food sovereignty · Gender · Critical pedagogy

Abbreviations

HIV	Human immunodeficiency virus
NGO	Non-governmental organization
PAR	Participatory action research
SFHC	Soils, Food and Healthy Communities organization

✉ Rachel Bezner Kerr
rbeznerkerr@cornell.edu

Extended author information available on the last page of the article

Introduction

Food sovereignty is a long way from lived reality in southern and eastern Africa. Chronic food insecurity and malnutrition are persistent problems for smallholder farming households in rural Tanzania and Malawi; two-thirds of households in Tanzania and at least one-third in Malawi experience food insecurity annually (Ellis and Manda 2012; Knueppel et al. 2010; National Bureau of Statistics and ICF Macro 2017). Key reasons for this food insecurity include poverty, persistent inequality and marginalization in the political system for smallholder farmers, and severe land degradation in southern and eastern Africa, including high rates of deforestation and soil degradation (Kangalawe et al. 2008; Zulu 2010). Further, gender inequality also contributes to food insecurity and malnutrition. Women do approximately half of all agricultural labor and contribute significantly to income generation, as well carrying out the majority of food processing, childcare and domestic tasks in both countries. However, they have little control over agricultural or nutritional decisions, and fewer agricultural extension and training opportunities (Bezner Kerr 2005; Peterman 2011). Inadequate nutritional knowledge about optimal child care and feeding practices, such as dietary diversity, frequent feeding and exclusive breastfeeding, also play an important role in these problems (National Statistical Office and ICF Macro 2017). In the face of climate change, this region is predicted to have higher temperatures, lower overall precipitation rates, and greater unpredictability of rainfall in general, resulting in reduced crop yields (Funk et al. 2008).

Intensification is the dominant model of agriculture promoted in Africa to address food insecurity and malnutrition, but has had limited success, with benefits mainly accruing to large landholders and input suppliers. In Malawi, for example, where there have been fertilizer and hybrid maize seed subsidies for over a decade, evidence suggests that wealthier large-scale farmers benefited from the subsidy (Chinsinga 2011 Chirwa and Dorward 2013), crop productivity is declining (Messina et al. 2017) while food insecurity and malnutrition remain high (National Statistical Office and ICF Macro 2017). As such, many argue that intensification is not a sustainable approach. A United Nations mission to Malawi in 2013, reflecting on the failure of an input-intensive approach to address food security, poverty and nutrition, called for more investment in farmer-led initiatives that used agroecological and participatory approaches (De Schutter 2013).

Agroecology is defined here as a holistic approach to agri-food systems, which uses ecological concepts, takes social sciences, indigenous and local knowledge systems into account, considers the broader political-economic aspects of agriculture and food, and is action-oriented, aiming to build a sustainable and equitable food system (Altieri 1995; Méndez et al. 2013). Agroecological approaches include agroforestry, crop diversification, cover crops, legume integration and organic production methods. These methods can reduce costs, build soil health and improve long-term community resilience to climate change and environmental and economic challenges (Snapp et al. 2010; Lin 2011; Wolfe 2013).

Because agroecology is not a one-size-fits-all approach to farming, it requires tailored understandings of climactic, biogeochemical and plant relationships. Agroecology educational approaches that link to food sovereignty also incorporate education on social inequalities that have marginalized indigenous and poor agrarian communities (Meek et al. 2017; McCune et al. 2016; Rosset et al. 2011). Industrial approaches to agriculture have downplayed and denigrated local indigenous knowledge in southern and eastern Africa for decades, often in concert with colonial and postcolonial discourses that emphasize the superiority of Western science and urban elites over African rural people's knowledge (Mulwafu 2015; Bezner Kerr 2014). Historically, agricultural extension relied on hierarchical forms of education (Chowa et al. 2013). A similar model has existed for health education; in most communities, community health workers have become a tool for one-way information transmission from academia and health experts rather than listening to the needs and problems of rural people. As such, there must be support for reinvigorating local knowledge and rebuilding farmer capacity if integrated agroecology strategies are to be implemented (Drinkwater and Snapp 2008; Bezner Kerr et al. 2018). This gap has led to calls for investment in agroecologically-sound extension and related adult education initiatives along with biodiversity and agroecology practices (De Schutter 2013; Snapp et al. 2010). At the same time, there is strong evidence that without investment in nutrition education, agricultural interventions are unlikely to have positive effects on maternal health and child nutrition (Arimond et al. 2010; Berti et al. 2004). Indeed, social inequalities at the household and community levels further exacerbate both food insecurity and malnutrition (Qureshi et al. 2015).

Although some educational material on agroecology has been developed, some assume an advanced formal education level of participants. For example, food sovereignty materials from Latin America focus on "agroecology as praxis" in which sophisticated and critical assessments of the political and economic foundations of the agri-food system are emphasized (McCune et al. 2014, 2016, 2017; Meek 2015). These approaches, while effective at building social movements in the context of Latin America, tend to be aimed at those with a higher level of literacy than is found in rural southern and eastern Africa.

Current agroecology materials are also context-specific, indeed a key feature of critical food system pedagogical approaches (Meek et al. 2017). Many of these educational efforts are in very different political, social and environmental contexts. While there are a host of food sovereignty-based agroecology educational programs in Latin America linked to social movements (e.g. McCune et al. 2016, 2017; Meek 2015; Meek et al. 2017), and food justice programs which use popular education methods to draw links between structural inequalities and agroecological practices in North America (Reynolds and Cohen 2016; Meek et al. 2017) in Sub-Saharan Africa there are few such examples. The few educational materials that do exist for Sub-Saharan Africa provide relevant details on practices, such as compost making, the importance of diverse diets, or participatory approaches to discussing gender-based violence, but do not explain the relationships and synergies between agroecology, food systems, health and social inequalities. Farmer field school approaches and extension reform movements are emerging in East Africa in support of farmer-led action-learning and in some cases pay attention to gender and power relations along with agroecological concepts (Morrone 2017). We were unable to find examples of farmer field school curriculum that supported integration to the extent here.

In Malawi, some of the authors have been engaged in long-term, participatory and collaborative research aimed at addressing food insecurity, malnutrition, climate change, land degradation, and social inequalities (Bezner Kerr et al. 2011, 2012, 2016). Through our collaborative work there, we have identified key strategies to address these issues and work towards food sovereignty (Bezner Kerr et al. 2014; Msachi et al. 2009; Snapp et al. 2010). These strategies, however, have not been formalized.

Theoretical underpinnings

This project wove together educational strategies from a number of disciplinary backgrounds, including participatory action research, feminist theory, transdisciplinary, several pedagogical methods: experiential learning, theatre for development and story-telling.

Participatory action research

PAR was an important guiding approach in the development of this curriculum. PAR is an experiential methodology in which poor, oppressed, exploited groups and social classes do research in order to transform their situation (Fals-Borda and Rahman 1991). We therefore strove to weave in iterative and experiential learning into the curriculum. Participatory education also pays attention to empowerment and power relations (Freire and Ramos 1970). Use of local experience and indigenous knowledge is part of action learning, since community members possess knowledge and cultural context (Freire and Ramos 1970). Scientific knowledge and education can be viewed as problematic, especially when knowledge is monopolized by experts. One core principle used by participatory researchers to address this tension is that of mutual respect, and building trust with practitioners who will use the knowledge (Snapp and Heong 2003).

Feminist theory

There are many epistemological strands that link PAR to feminist scholarship. Feminist praxis includes several approaches to address inequalities as outlined by Naples (2013): strategies for inclusion, methods of empowerment, countering power imbalances, organizing across differences and reflexivity. These educational approaches include creating spaces that encourage those who may have limited political clout and power to have discussions, share ideas, resolve conflicts or raise concerns, giving opportunities for more marginalized groups to negotiate, contest and overcome inequalities (Hassim 2009). Feminist standpoint theory posits that those who are marginalized and oppressed have knowledge and experience that can help to solve their own problems, shape knowledge and understanding (Harding 1986; Hartsock 1999; Collins 1991). Educational methods informed by feminist theory encourage collective conversations, dialogue, story-telling, problem-solving and drawing from lived experiences.

Transdisciplinarity

Transdisciplinary approaches are defined as those which integrate different knowledge, not only from a range of different disciplines, but also from different knowledge systems (e.g. local, indigenous, experimental, historical), and involve a diverse set of stakeholders in generating new knowledge. A transdisciplinary approach is also problem-oriented, rather than focused on generating novel information for its own sake (Méndez et al. 2013).

Objective

Therefore, we set out to create a curriculum that would synthesize smallholder farmer knowledge, crystallize best practices in terms of agroecology, nutrition, climate change and social equity, and convey the interconnectedness of these domains. The intention was to develop a publicly available resource,[1] that others can use and adapt in other settings. Specifically, in this paper, we describe the process of the development of this curriculum and experiences with its first round of implementation and briefly discuss these results in the context of recent literature on critical food system pedagogy, feminist praxis and food sovereignty.

Didactic methodologies, study case and methods

Didactic methodologies

In light of this theoretical underpinning, we used three educational strategies for learning: experiential learning, drama and small group discussions. These three strategies allowed us to draw on an epistemological stance that emphasizes respect for local knowledge, recognition of the

[1] See http://soilandfood.org/projects/participatory-integrated-curriculum-project/.

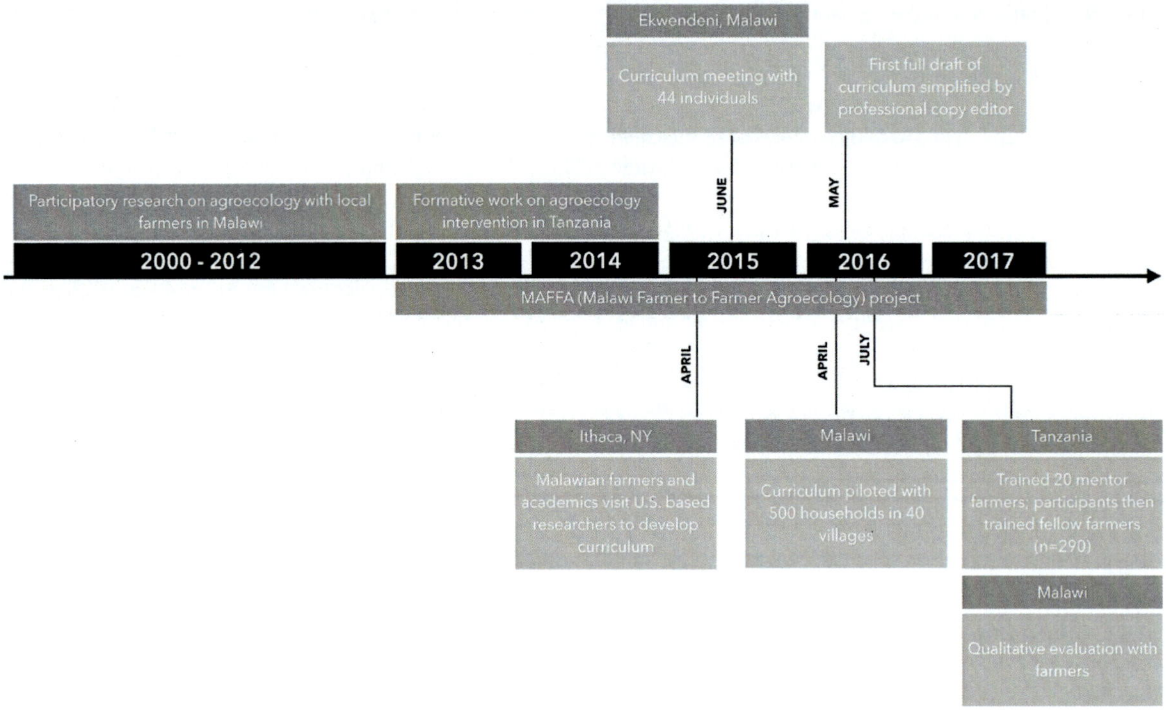

Fig. 1 Timeline of curriculum development process

interconnectedness of ecological, social and health issues, and transformational approaches to education.

Experiential learning

Put simply, experiential learning is learning by doing. There is much to support its effectiveness. We therefore strove to weave in iterative and experiential learning into the curriculum. Experiential learning is at the basis of participatory action research in agroecology, and is foundational for many critical food system programs in agroecology in the Americas (Meek et al. 2017).

Theater for development

The use of theatre and performance as a form of critical pedagogy has been central in Sub-Saharan Africa since the mid-1970s (Kerr 1995). Since then universities across Africa adopted it to work with communities. Theatre for development includes dialogue, participatory methods of theatre of the oppressed and communities creating performances. Theatre is now a common tool of engaging communities in social issues, involving them in debates around the nature, effects of the problems on them, and finding solutions to resolve them (Sloman 2012). Theatre has been used in health, education, governance, agriculture, gender equity projects around the world (Sloman 2012).

Small group discussions

Discussions in small groups encourage active learning processes by sharing experience, knowledge, reflection and group problem solving (Abusabha et al. 1999; Affleck and Pelto 2012). In critical food systems pedagogy, this reflection and discussion is crucial for considering ways to transform the food system (Meek et al. 2017).

Study case: curriculum development activities

The curriculum that emerged was the result of more than a decade of work (Fig. 1). The beginnings of the curriculum were in northern Malawi in 2000. There, transdisciplinary and participatory action research has been carried out by a team of smallholder farmers, agricultural scientists, sociologists, nutritionists, geographers, medical and NGO staff (Bezner Kerr et al. 2011, 2012, 2016; Nyantakyi-Frimpong et al. 2017). Team members have been engaged in long-term, participatory and collaborative research aimed at addressing food insecurity, undernutrition, climate change, land degradation and social inequalities. They have shown that

Table 1 Individuals who contributed to the development of the curriculum

Institution/Profession	Area of expertise	Number of individuals	Number of women
University researcher	Social science	4	2
University researcher	Nutrition	2	2
University researcher	Agroecology/Soils	5	3
University researcher	Extension	2	2
NGO staff	Community development, agroecology, participatory methods	2	1
Hospital staff	Maternal and child nutrition, HIV/AIDS, community development, youth issues	8	4
Farmer	Agroecology, nutrition, social issues	16	6
Youth group leaders/farmers	Youth concerns, farm and food enterprises, HIV/AIDS	5	2
Total		44	22

participatory, farmer-led agroecological experimentation can build farmer knowledge, foster innovation, and have demonstrable impacts on food security, nutrition and sustainable land management (Bezner Kerr et al. 2007, 2014; Snapp et al. 2010). In 2012, they began to scale out this work, by training 6000 farming households on agroecology, social equity and nutrition issues (Nyantakyi-Frimpong et al. 2017).

In 2013, we decided that it would be useful to formalize the participatory techniques used in Malawi, in part because of our on-going participatory research on climate change adaptation. We realized the need to carry out discussions and education about climate change integrated with agroecology and social equity, but we lacked the educational tools to do so (Bezner Kerr et al. 2018). As such, we reviewed the literature to identify curricula related to the key topics: agroecology, social equity, nutrition, and climate change. This review was not a comprehensive overview, but rather an examination of the 'grey' and published curricula that could be obtained through internet and library searches and through correspondence with people in relevant fields. We found that although there were some curricula in existence (e.g Menza and Probart 2013; Okoth and Nalyongo 2013) there were few materials to draw on that were designed for people with limited formal education. Instead the focus of curricula identified was on students and facilitators with formal educational backgrounds, often using language, jargon and abstract concepts that presumed prior training in this subject area. A second shortcoming identified was that few curricula integrated across topics; most focused on just one or two areas (e.g. agroecology and nutrition (see for example MAIWD 2015) or gender equity and infant and child nutrition (e.g. RWANMREC 2014).

After 4 months of discussions and preliminary material gathering, in April 2015 six Malawian team members, including 4 farmers, 1 academic and 1 NGO staff member traveled to the United States and held in-person meetings (Fig. 1). After this preliminary discussion, follow-up meetings took place by Skype. Farmers in Malawi were actively participating in these meetings to develop the curriculum. Different team members were assigned different sections of the curriculum to write, with a goal to have drafts ready for 'testing' in June of that year.

In June 2015, 44 members of the research team met in Ekwendeni, Malawi for a week to review the materials developed, revise and assimilate them (Table 1). Twenty Malawian fårmers were included in the meeting. The farmers were purposively selected to include those with extensive expertise in agroecological methods, nutrition education, climate change, or having little to no exposure to any of these ideas. There were also several youth members from a drama group. The farmers came from a range of levels of food security and ages. In the week we were together, we worked on the curriculum content and delivery, had field trips and enjoyed social activities together.

With help from multiple team members who submitted sections of the curriculum, the lead author assimilated materials from the June 2015 meeting into a single document that could then be piloted in Malawi with rural communities. A draft version was available in October 2015 and translated into two different languages (Tumbuka and Chewa). Illustrative materials were developed concurrently. The full draft of the curriculum in these two languages was ready by March 2016. In April and July 2016, a sample of 500 Malawian households from across 40 villages participated in a 2-week training; 200 households were in the Mzimba District (northern Malawi) and 300 in Dedza District (central Malawi). These two regions were selected because they both have high levels of food insecurity and undernutrition, but differ both in their agroecosystems, including rainfall and cropping patterns, and different land tenure systems which influence gender relations, providing for interesting comparisons. Each training day consisted of morning and afternoon lessons that cut across all curriculum topics. Based

Table 2 Gender, location and age of qualitative interview respondents (N = 38)

	Total	Drama	Non-drama
Region			
Mzimba	20	13	7
Deda	18	17	1
Gender			
Male	15	12	3
Female	23	18	5
Age			
18–33 years	14	11	3
34–49 years	16	13	3
50–65 years	6	4	2
66 + years	2	2	0

on feedback and experiences in April 2016, the curriculum was revised and finalized (language simplified and harmonized) by a professional copy editor and the lead author, and then re-translated into three languages: Swahili, Chewa and Tumbuka.

Methods: curriculum assessment

In July 2016, a qualitative evaluation was conducted in Malawi to evaluate experiences with the curriculum. Thirty-eight qualitative interviews and observations were conducted by a Malawian staff person and an American communications PhD student, to help the research team better understand farmers' engagement in the training overall. Respondents discussed their experiences for approximately 1–1.5 h with the 2 research team members in Chichewa and Tumbuka. Interviews touched on perceived value and challenges to the curriculum, engagement, comprehension of training, attitudes, self- and collective efficacy and information sharing after training. The interview respondents were selected randomly from the complete list of 500 curriculum respondents. Purposive selection of the list was done to ensure that men, women, youth and elderly participants were all interviewed (Table 2). Interviews took place at the two training locations (local buildings used for church and community meetings) or farmers' homes.

Interviews were transcribed and translated into English. They were then coded using ATLAS.ti for qualitative data analysis. The analysis was structured around 7 code families: overall perception of value, overall perception of challenges, social interaction and information sharing during training, comprehension, social interaction and information sharing after training, efficacy, and attitudes. Over a dozen sub-codes helped the team further analyze the responses in each of these code families. The codes were developed using inductive, grounded theory.

Firsthand observations were also carried out during the curriculum training by four Malawian researchers and the American PhD student. In addition, follow-up observations were made by Malawian researchers during follow-up activities in the communities several months after the training took place. These observations were used to validate the initial findings drawn from the interview data.

The research was approved by the Cornell University Institutional Review Board (Protocol # 1507005688) for human subjects research in the United States and the National Commission for Science and Technology, National Committee for Research in the Social Sciences and Humanities (Protocol #11/15/63) in Malawi.

Results

Our results include a description of the curriculum development process and initial experiences with this integrated, participatory curriculum in Malawi.

Curriculum development

Initially the team had regular web-based discussions to discuss how to organize the curriculum. These discussions proved very challenging, as there were highly divergent views about how to develop the material, what topics needed to be included, and how to write the material. While there had been considerable exchange and preparation before the face-to-face meeting, there was little forward progress or common understanding between the diverse perspectives until everyone was 'in the field' and able to discuss and reflect together in Malawi. This finding speaks to the importance of experiential learning—even in the curriculum development itself. Mutual respect for the multiple ways of knowing about farming was an essential aspect of our approach. We saw as central to this effort the trust and quality relationships that were built up over years of interaction between a farmer researcher group in Northern Malawi and some of the academics who initiated this curriculum (Bezner Kerr et al. 2007; Nyantakyi-Frimpong et al. 2017). Trust takes time to develop, and continuing commitment to engagement and communication is an essential process sometimes overlooked in participatory action research and education efforts (Snapp and Heong 2003). One key dimension of building relationships and trust in this curriculum were the field trips and opportunities for face-to-face dialogues. One week in person was just the start of breaking down barriers and identifying diverse strengths.

Participatory action research aims to break down hierarchical ways of knowing. Academics shared knowledge of processes difficult to observe and sometimes slow to change, such as soil biology and organic matter transformations.

Farmers appreciated learning more about the connections between different parts of the agroecosystem. The participating farmers, once recognizing that *their* way was being respected as real knowledge, began to give additional examples of connections between, for example, trees and water conservation, chickens and pest control, plant diversity and greater nutritional security. Therefore, taking an agroecology approach to curriculum development meant that ideas coming from the academics dove-tailed with the smallholder experience.

A key debate amongst the curriculum team was how to integrate these four primary topics (agroecology, nutrition, climate change and social equity) in ways that were meaningful. One key idea that arose from the face-to-face meeting was to create dramas that would help to integrate the different topics together. These dramas would be performed daily by curriculum participants using theatre methodologies, as a means to foster critical engagement with the topics and promote transformative change. A small group of graduate, undergraduate students and a nutritional anthropologist worked to develop these dramatic stories that threaded together the different dimensions of the curriculum (Box 1). Two methods were used with the dramas. Firstly, using local names, and typical agricultural, nutritional and social problems faced by the farmers, the stories became blueprints for the created dramas to be shown to all participants during plenary sessions in order to provoke participatory dialogue. The participation of the farmers in the creation of the dramas, and the debate that followed during plenary sessions brought the problems right in front of the community where they were analysed and possible solutions proposed. In the second approach, stories had to be created from the training sessions' participatory discussion of problems raised in the different agricultural, nutrition, climate change and gender equity issues. Participants were then expected to create dramas around the issues with guidance from participatory drama facilitators, animating for example, ways in which pests attack crops and ways in which this can be dealt with using agro-ecological approaches such as crop diversity, rotation, intercropping etc. without use of pesticides, which can have negative health, economic impacts alongside biodiversity impacts. Shown to other participants during plenary sessions, this approach exhibited novel creativity from the different groups in story creation, and innovative performance, inclusive of local performance forms. The two approaches made the curriculum commonplace for the participating farmers and made the material more understandable, and applicable to their local experience.

Integrating critical food system pedagogies into the curriculum in ways that challenged the current political order and fostered food sovereignty was also a major challenge. One way that the team attempted to do so was by incorporating stories as a teaching tool (Box 2). These stories were provided by local farmers and during testing were adjusted to reflect local political realities.

Field trips done during curriculum development facilitated dialogue across disciplinary, educational and cultural divides, and allowed the farmers to use their indigenous knowledge bank as a starting point for further education and exploration. Farmer trainers enjoyed the exercise of going to a field, and then deconstructing what they saw from an agroecological perspective, elaborating on the processes that farmers were using. The field trip also provided stimulus for many of the team members to base the curriculum in experiential, transformational and participatory action research approaches, as it reinforced poor farmers' abilities to carry out their own research, to assess their problems, and the integrated nature of the problems of land degradation, food insecurity social inequalities, climate change and malnutrition (see Box 3). In the case of Mrs. Tembo, for example, her training in agroecology has provided her with a greater recognition about how things are connected—how they are ecological. In some ways she has shifted her farming systems from a game of checkers to one of chess—where she is required to think several steps ahead. This approach may make it look like extra work to her neighbors, but it is in fact more of a "walking alongside" the system. Her experiences have also shifted her family's nutritional and social circumstances, demonstrating the integrated nature of these issues. These cases provided useful examples of ways that farmers integrate nutrition, agroecology and social equity for the full team to try to apply to the curriculum material.

There was considerable experience within the team using inquiry-based, experiential and participatory learning strategies, such as activities to demonstrate topics and explore in a co-learning manner. Field observations are another approach that is widely used in Farmer Field School and related action-learning approaches (SUSTAINET 2010). However, many team members were largely familiar with conventional approaches that emphasized expert scientific knowledge delivered through group presentations with visual learning aids. Thus, how the curriculum could be 'delivered' in transformational ways became a focal point for debate and discussion. Key ideas from social learning (e.g. multidimensional learning) and critical pedagogies (e.g. horizontal learning, reflexivity) were discussed and emphasized by some members of the team. Relatedly, there were questions about whether there should be a focus on specific topics or on methods of learning. There were also questions about the format and visual materials that would make up the curriculum. Finally there were questions about assessment. How would we be able to assess success of the curriculum? In addition, many of the materials identified were not useable for those with limited formal education, which worked against the vision of a 'farmer-led' curriculum. One common challenge was how to write the curriculum in such a way to make the material accessible for those with less than a secondary education level in rural Malawi and Tanzania. Much effort was made both in terms of language accessibility, visual materials and hands-on experiential learning to make the curriculum truly accessible and useable by farmers.

Shared language was an ongoing struggle during curriculum development. The climate change module was particularly challenging to develop because of the abstract and scientifically technical terms needed to grasp some of the concepts. Many hours of discussion between farmers and scientists about concepts and approaches led to revised material. The curriculum begins with engaging farmers in an exercise to characterize historical weather patterns (e.g., rainfall, temperature) for their region in relation to a wide range of farm decisions (e.g. crop selection, soil and water management, arrival of insect pests). The concept of climate change was then introduced by asking them to identify changes in weather patterns and extreme weather events compared to what previous generations of farmers in their region have had to face. As part of the curriculum they were asked to prioritize the most critical weather-related challenges they face today, and to identify successful approaches for adapting to these challenges. This component of the curriculum builds on a framework of ecologically based and local-knowledge intensive approaches to climate change adaptation, and efforts to draw out these linkages were a key aspect of this part of the curriculum. Crop diversification and building soil organic matter builds resilience to weather risks, particularly drought and flooding. Additional modules introduce farmers to some more complex information about weather patterns and climate change—such as the relevance of El Nino and La Nina cycles to seasonal weather patterns where they live, the future projections of climate change for their region, and the uncertainties in these projections. A final phase of the climate change module introduced greenhouse gases, and farming practices that can help to slow the pace of climate change, such as storing carbon in trees and soils. Efforts were made to link climate change adaptation to other dimensions of rural life, for example, improving soil health not only helps with long-term food security, but is both a climate change adaptation and mitigation strategy.

Shared language was much less of a challenge in the Agroecology module. Over a decade of collaboration on participatory action research among SFHC members meant that there was a shared history of working together on agroecology for a number of the farmers and scientists present. This helped build mutual understanding of terms such as soil health, rotations and beneficial pests. It took some time to confirm shared understanding of terms, such as 'agroecology' which some farmers called good farming or natural farming, yet there was much commonality in experience with observations being made such as soil moisture being high under mulch, or insect diversity in mixed plantings that included perennials or legume species. The farm visits with SFHC members contributed significantly to success in this module. Having farmers explain their farming systems to all curriculum contributors allowed organization of this module based on what actually happens on an agroecological farm. Based on the field visit experience, the agroecology modules began with observational activities of key features of 'farming with nature.' Crop diversity, which was a key practice of many experienced farmers, was the second module as part of a focus on agrobiodiversity more broadly. Soil health was included in two separate modules, since the team felt that understanding basic principles of soil health followed by practical modules on how to improve soil with organic material was needed. Soil and water conservation made up an additional two modules, building on previous experiences, followed by two modules on weed, insect and disease management. Finally there was a module focused on planning a farm which drew on ecological principles.

The development of the nutrition module took as a starting point curricula designed by UNICEF (2013) and FAO (Okoth and Nalyongo 2013; MAWID 2015) to be taught by community health volunteers and in farmer field schools. An outline based on these materials was shared with farmers who then made improvements based on their experiences. In general, farmers agreed that the topics covered were important, but encouraged more practical learning. For example, suggestions were made to include discussions on overcoming barriers to the "best practices" suggested and a module on healthy cooking practices. Moreover, farmers also suggested changes to the proposed learning activities. The flipcharts and markers that were initially proposed were considered impractical since such supplies are not often readily available and assume literacy of all participants. Instead, alternative styles of learning, e.g. legume cooking lessons, with legume recipes included in the curriculum, were augmented. In the final curriculum, the nutrition module covered the following topics: importance of dietary diversity for all member of households; different nutrition needs of pregnant mothers, infants, school-aged children, and adolescents; proper infant and young child feeding practices i.e. breastfeeding and complementary feeding; and "capstone" of a cooking session.

The social equity module was developed initially by discussing inequality issues based on the long-term experience of the curriculum team, and also drew on training modules that emphasized gender and other forms of social inequalities in health and child care (PATH and CARE 2011; RWANMREC 2014; Werner and Bower 1991). Farmers and NGO staff developed dramas, case studies and stories to illustrate the different types of inequalities based on their experiences. Specific sessions in the final version included an introductory sessions on inequality, gender inequality, gender roles, division of labor at home and other places, healthy relationships, family budgets and food, alcohol and drug abuse, gender inequality and HIV, positive parenting and two modules on gender-based violence. The final version of the curriculum also integrated lessons across the four themes of agroecology, social inequity, nutrition and climate change, and included sessions on participatory action research and approaches, how to carry out experiments.

Box 1: Example of a drama developed for the curriculum: Johnny and Mary

Instructions:

Assign roles for Johnny and Mary. Ask the actors to perform this story for the group.

Story: Johnny and Mary are starting their married life and working on their farm. They have a long talk about how the future is uncertain, both for their farm and for their family. They discuss how there has been very little rain so far, and they do not know when it will rain next. People in the village used to be able to count on rain to help their crops grow, but they do not know what will happen this year. They can sense that the climate is changing. Johnny and Mary also want to have their first baby, but they do not know when Mary will become pregnant. Thinking about these uncertainties makes them anxious. Johnny and Mary decide that the best solution is to be prepared for all possible outcomes. They discuss the practices that will help them capture and efficiently use whatever rain may come. They agree to save the money they received at their wedding until their baby is born, to help pay for the expenses.

Discuss

What changes can they make on the farm to adapt to dryer or wetter conditions?

How can they be more prepared for the baby?

Box 2: Sample of Story-telling in Curriculum that addressed Social Inequalities

Sample of Curriculum

Activity 3: Local politics and politics on a broader scale

This activity will help the group understand local inequalities and how political power works.

Duration: 45 min

Now we are going to discuss some of the challenges that poor people face as farmers all over Malawi and Tanzania. Tell the story below, or make up a similar one that reflects the local inequalities. Then ask the group what they think of the story. Can they think of stories or examples from their own lives when people have been affected by differences in power?

Story: A Rich Man in a Rural Area

A rich man came to the village and made a deal with the village headman. The headman sold a big plot of land to him, even though the land was being used by poor families that had worked this land for many years. The poor families were kicked off the land by police with guns, who said it is now someone else's land. When the poor families tried to talk to the headman, he said that they didn't have rights to the land, even though according to communal law and custom, they did have rights. Then the rich people came and started an estate to grow macadamia nuts, coffee, tobacco, and other export crops. They tried to talk to the rich man, but he simply arrested them and increased the security guards on the land.

Discuss

- Does this kind of thing happen in your communities? Can you share a story about it?
- What are the later effects for the people involved?
- What can local people do to help address this situation?

In some parts of Malawi and Tanzania, poor people have organized to fight peacefully for their rights. They have gone to the courts to demand their land back, or they protest peacefully by occupying the land and demanding access to it.

Box 3: Field Trip Case Studies

Case 1: Mrs. Tembo

Mrs. Tembo is married with 5 children. She grows pigeon peas intercropped with groundnuts. She began experimenting and being part of SFHC in 2012. She has bought 2 goats from the proceeds of her legume crops. She is improving the soils, the family foods, and nutrition. Before being part of SFHC, her children were often sick. Now they are not, which she thinks is because of the dietary diversity brought by growing pigeon peas. She will rattoon the pigeonpea and plant maize here next year. Some of her neighbors have adopted agroecology practices, and come to her farm to see what she does, while others are skeptical and have not followed suit. At first she found these methods were a lot more work, but now she just sees it as how things are done. Her husband used to see groundnuts as a woman's crop, and let her do all the work. After the first harvest, she kept all the money from selling the crop, would not share it with her husband when he asked, because she had done all the work. The next planting season, her husband told her that he thought she needed his help, and since then has worked with her on agroecological methods.

Case 2: Mr. Mvula

Mr. Mvula is married with 4 children, now all grown. He started using these methods in 2006, after learning about them from SFHC. He has focused on agroforestry tree species such as *Gliricidia* which help build soil quality and protect the fields from erosion. Msangu sangu also planted to improve soils. The Mvula family chose what to plant based on what would best feed their family. When he compares his crops and fields to his neighbors, his fields stay moist significantly longer after rains because of the shade from shrubs and trees. The work is shared between Mr. and Mrs. Mvula, he bragged and his wife admitted that he is even a good cook!

Curriculum content

In this section we will outline the curriculum content, which included 4 subject areas: agroecology, climate change, nutrition and social equity. Transformative learning approaches which integrate participatory action research, transdisciplinarity and feminist praxis are used throughout the curriculum.

Agroecology

Many of the agroecology sections of the curriculum involved hands-on experiential activities including close study of soils, insects and farming systems. Key principles were introduced through these activities. Since smallholder farmers have daily experience with the local agroecosystem, that surrounds them, they can quite easily think and act in agroecological terms. An iterative and experiential learning process to integrate this knowledge is at the foundation of participatory action research in agroecology, namely: plan, action, reflect and iteratively learn together (Moncure and Francis 2011; Méndez et al. 2013). The curriculum attempted to weave in this reflection and observation as part of the pedagogical approach.

Climate change

Climate change exacerbates food security vulnerabilities of the Global South, and strengthening anticipatory capacity for this new challenge is essential (Tschakert and Dietrich 2010). Numerous studies have documented the importance of indigenous ecological knowledge for elucidating complex climate and environmental trends, and for developing adaptation strategies that are feasible, acceptable, and effective (Tschakert and Dietrich 2010; Mistry and Berardi 2016; Popkin 2016). Integration of local observations with scientific data provides farmers with additional input to help them discern between "normal" weather variability and a shift in weather patterns warranting capital investment or a change in farming practices (Wolfe 2013). At the same time, bringing in indigenous knowledge and perceptions is essential for bridging the gap between scientific discourse and reality 'on the ground' for people experiencing the impacts of climate change (Bezner Kerr et al. 2018).

Nutrition

While agroecological approaches have been shown to have positive impacts on ecosystems and climate change adaptation, there is less evidence that they have a direct impact on nutrition and food security. This education approach focused on the linkages between nutrition and agriculture through participatory education strategies. There are few examples, where agriculture and nutrition education are explicitly integrated into a participatory action educational strategy, and even fewer that use an agroecological foundation. There is strong evidence that without investment in nutrition education, and attention to social inequalities at the household and community level, no agricultural intervention is likely to have positive effects on maternal health and child nutrition (Arimond et al. 2010; Berti et al. 2004). One key reason for this lack of effect has to do with issues of social inequality at the household and community level. Furthermore, structural inequalities at the national and international level shape the possibilities for smallholders to make transformative change within the food system and—hence the need to incorporate critical food system pedagogical approaches that address social inequalities at multiple scales (Meek et al. 2017).

Social equity

Participatory action research has a core focus on social equity concerns, such as class inequalities and the research approach often includes a notion of *conscientization*-raising awareness about social injustices. This curriculum focused on some of these dimensions and sought to raise awareness about broader political and economic inequalities which prevent food sovereignty. Severe poverty, class inequalities and limited educational opportunities restrict smallholder farmers' ability to step back from their specific situation and carry out a broader analysis of their oppression. In addition, gender inequities across multiple scales are a pervasive structural issue–within households, kin networks, farmer organizations, extension agents, national and international policies, which in turn affect food security and nutrition outcomes. In addition, there are many ways in which relations and processes between agriculture, gender and social dynamics and nutrition can intersect (see for example Bezner Kerr 2017; Bezner Kerr et al. 2008; Berti et al. 2004). This curriculum was an attempt to bring these linkages to the forefront.

Curriculum implementation

While integrating the four topics together was one way to ensure that questions of equity can be discussed, this strategy could be controversial. In Tanzania, one participant rejected the gender and family planning topics in the curriculum, insisting that he was only interested in agriculture topics. When the group discussed family planning, he refused to participate. The facilitator used this an opportunity to have an in-depth discussion with the participants about this objection, but the initial participant left the training. After he did so, the team discussed with the rest of participants what they thought and what steps could be taken to address similar objections in the future. The rest of the participants discussed the different resources they already have in the villages, their views, and on the possibility of discussing this issue with local religious leaders. A follow-up visit was made to the village where this participant resided, and discussions held including with village leadership. The consensus was that another village member be nominated. This experience opened up the discussion about gender and family planning more broadly, and this village is currently one of the more active in the Tanzania project.

Nearly all of the farmers interviewed described the drama, stories, experiential activities and small group discussions as helping them and others comprehend the material. Farmers also said that the light-hearted tone of the drama made it easier for them to talk about sensitive subjects, such as HIV/AIDS, alcoholism, and violence in the household. Small group discussions were also described as helping farmers review and clarify information. Drama added the benefit of a visual demonstration, which farmers described as central to helping them understand complex concepts, or of making difficult topics more accessible. They explained that much is missed in the classroom setting because everyone has different education level and learns differently, but believed that everyone can understand the material when put into a drama. Drama also allowed the intersection of topics, such as climate, the farm, and household, to be depicted.

In Tanzania, there was some resistance to the participatory methods from some participants, who considered the use of drama and songs childish, while teaching in a speech was seen as more dignified. Some participants also complained that drama took a long time and required preparation and props. Facilitators asked the participants to see this training as an opportunity to gain skills to teach using drama. The participants are then free to use it or not when teaching in villages. The team and participants also discussed the possibility of using story telling as a simpler (logistically) and less 'embarrassing' alternative to drama. There were some facilitators who did not take participatory approaches seriously. This issue points to the challenge of creating a curriculum which challenges power hierarchies at multiple scales, since there may be less acceptance of participatory, transdisciplinary approaches by those who hold some vested power.

In Malawi, there was less hesitation about the use of drama and experiential learning, and the farmers' participation confirmed the approach. During the training, some of the topics were complemented with traditional songs that either indicated agreement with the topic being discussed or just to encourage the facilitator to continue with the 'helpful' discussion. On pests and ecological pest control in which birds become helpful in eating such pests as grasshoppers and worms, for example, the song *'kwa kwa kwa aliranji makwangwala'*–(*kwa kwa kwa* is the imitation of sounds made by crows)- and the other part translates–*why are crows crowing?* was sung. Elderly people willingly took part in the curriculum experiential and theatre activities. One 70-year old village chief offered to play the part of chief in a drama that showed well-to-do people and chiefs manipulating young poor women for sex in return for money. In another example, a 68-year old man played an insect that came to attack crops, and was then later eaten by a bird, played by an even older woman. Some of the drama facilitators had had previous experience with the elderly actively participating in drama in other field workshops in Mozambique.

As a pedagogical tool that aimed to draw out indigenous farmer knowledge, generate discussion and foster transformational change towards food sovereignty, there was some evidence of success. Around half of the farmers interviewed felt that the drama, story-telling and small group discussions gave them an opportunity to share their own experiences and innovations around curriculum topics, and for others to do the same, increasing overall knowledge sharing. Many of the participants reported sharing information from the training with other farmers through the use of drama, with some having traveled to up to 12 villages, and others reported being invited to come to neighboring villages that had learned about the teaching and dramas. When going to other villages, these farmers said they would notify the village headman that they were coming and maybe 50–100 people would be gathered; but once they started the drama, up to 200 more people would come to watch. A female farmer in her mid-30s living in Northern Malawi with a three acre farm and family of three described going to six villages with her drama group to share what they learned in the training.

> The vice chair or chair of the group would explain point by point and everyone would listen; after that they would perform the drama. The village head would let people know and they would gather, but when they started the drama more men, women, and children would come –'the drama invites people by itself.' When you tell people there are plays about something,

everyone would love to come –it's a good way to meet the majority of people. (Interview #3 6/21/16)

During interviews, people gave examples of attitude shifts and comprehension of new ideas related to both gender issues and climate change. A 37-year old male farmer with a family of 4 said: "When you explain to people about cutting trees, making charcoal, and gas emissions people might not understand it, but in drama the meaning comes to them." He added,

> When people are acting out the drama, you pick the things from the act and realize that all along I thought if I was doing this to my wife, I was doing the right thing, but from what you see in the drama you realize you were wrong all along. You might thing polygamy is ok because of your background, but when others explain and act out a drama you realize all along you were believing the wrong things. (Interview #2, 6/20/16)

While this respondent may have been saying what he thought the interviewers wanted to hear, observations by local staff in the communities in the following months indicated that farmers were using the curriculum as a regular resource to consult when they had questions about farming, nutrition or other topics. Participating farmers were also observed to share what they had learned with other people in their communities through both drama and discussions, on subjects such as soil types, gender and child nutrition. The curriculum pedagogical approach appeared to foster farmers' increased confidence in their knowledge and teaching abilities—evidence of potential for transformational change. Many farmers described a shift from being very shy, and never speaking in front of others, to enjoying teaching others challenging content through both lessons and drama. Several women reported being very quiet at the beginning of the training, but by the end, they were often the ones leading the drama and the discussions. The women said that the men supported this change; and the men said that after learning about gender equality in the classroom, they saw more value in encouraging the women to speak up. Dramas, stories and small group discussions allowed farmers to incorporate their own experiences and talk about how to overcome challenges together.

The elderly had the most trouble participating, but would often be given simple roles, and would participate in the discussion portion. A 62-year old female farmer described the mix of people in her breakout group of 15. She said there was "a mix of men and women with 4 elders." She describes the elders as "taking part, but at first they were feeling shy, but later they came in and started participating." The format of the curriculum had allowed her to share with others her knowledge of caring for children, among other things, and helped her feel more confident about sharing what she had learned during the training. She said that if there was a group of women, "I would be willing to teach them some new cooking methods, I would gladly do that," despite describing herself as shy before the training. She went on to say that "For women, at first they were shy, but later they opened up and started participating."

One woman farmer described one of the valuable lessons that she had learned during the training: "With the traditional way of farming, which was planting far apart (ridges and planting stations), they would only plant one crop in a field, but now [we have] learned to minimize the distance between the ridges and plant different crops in one field. Now depending on the rains, if one fails another may do well."

Discussion

Our objectives were to describe the process of the development and initial implementation of an integrated and participatory curriculum about agroecology, climate change, nutrition, and social equity that would be appropriate for smallholder farmers in sub-Saharan Africa. Although preliminary, these results suggest that the curriculum can be effective. More quantitative and qualitative data on the impact of the curriculum on knowledge, behaviors and practices will provide insights regarding how the curriculum has been useful, and at the same time identify gaps and how it needs modification. Measurement of impacts on agricultural practices, social equity, food security, and nutritional status will inform us if the curriculum can work towards transformational change. The curriculum development was not easy; it was a process of many negotiations—technical, cultural, and didactic.

The experience thus far in Malawi and Tanzania suggests that this transdisciplinary, participatory approach can support meaningful educational tools on agroecology, climate change, nutrition and social equity that foster knowledge shifts in multiple arenas, and that cross-fertilize to encourage farmer teaching within and beyond their communities. The curriculum was developed within the particular context found for smallholders in Malawi and Tanzania, and as such would need to be adapted for other agroecosystems, political and socio-environmental contexts. Ensuring that the language used is straightforward, concrete and jargon-free is important, as is the use of teaching strategies that encourage reflection, discussion and active participation. The process of developing the curriculum was as critical as the topics and approaches used – face-to-face dialogue in the location where the curriculum could be used, opportunities for field visits, reflection and exchange, and ample time for discussion were all crucial. As well, involving scientists and community development workers already attune to participatory, transdisciplinary approaches, prior experience in the places where the curriculum was being developed, strong partnerships with rural communities, and openness to learning were important to success (Snapp and Heong 2003). Despite these foundations, there were obstacles to constructive dialogue, and a commitment to see the project to fruition proved critical to completion. The foundations of action research with-and-for marginalized groups was critical to ensure long-term commitment to completion.

Overall, the curriculum provided teaching tools that allowed for learning, open discussion and problem-solving - participants were able to discuss and come up with solutions to some challenges which they face in their communities. The material in the initial draft needed 'translation' by those facilitators familiar with the concepts, and required further revision to make it more accessible. The hands-on experience sessions added clarity of the issues, for example a group visit to the local bush made it easy to explain diversification of plants, or the soil health session, which included molding of different soils made it easy to clarify differences between sandy, clay, and loamy-sandy soils. The use of drama, discussion and songs created a festive and enjoyable atmosphere while fostering dialogue and learning.

We link these results to recent literature on critical food system pedagogy, feminist praxis and food sovereignty. The experience of this team in generating an integrated curriculum using participatory methods drew on concepts and methods from a range of disciplines, including participatory action research, experiential learning, critical food system pedagogies, transdisciplinary approaches and feminist praxis (Table 3). This experience has commonalities with recent findings in 6 different food sovereignty educational programs in the Americas, which were borne out of common themes of people's social inequality, food insecurity and marginalization from agricultural knowledge which is divorced from the social or political context (Meek et al. 2017). The strong emphasis on examining and questioning different forms of inequality in relation to other dimensions of rural life—farming, nutrition - is another common thread with recent critical pedagogical efforts on food sovereignty (Meek et al. 2017).

Similarly, participatory action research, dialogue and consciousness raising through hands-on experiences are all pedagogical tools used in these programs (Meek et al. 2017), as is the case with this curriculum. Playfulness and performativity, themselves often part of feminist praxis, can be important ways to animate people, and song and dramatic performance can then facilitate taking part in a forum that generates social change (Patel et al. 2015). Tools of small group discussion and theatre allowed marginalized groups to question and explore different dimensions of inequality—a key aspect of feminist praxis (Naples 2013).

One of the challenges faced in the curriculum was introducing potentially new forms of farming with agroecological practices while respecting traditional knowledge. Feminist standpoint theory emphasizes the relevance of local, subjective knowledge (Harding 1986) while participatory action research also values local experience and knowledge (Méndez et al. 2013). As some of the quotes above reveal, some farmers thought of certain practices such as monocrops as traditional even though historical research on Malawian agricultural practices indicates that traditional practices included agroecological methods such as intercropping (Mulwafu 2011; Moyo 2014). Active efforts by colonial and postcolonial governments encouraged monocropping, reliance on fertilizer and purchased seed, and minimized or even denigrated traditional knowledge including use of a diverse range of crops (Mulwafu 2011; Bezner Kerr 2014). Over many decades, some traditional practices have been lost, or forgotten, including indigenous crop varieties (Bezner Kerr 2014); recovering this knowledge and introducing new agroecological practices requires sensitivity and respect for local knowledge and recognition of some of these power imbalances and contradictions. This curriculum didn't always 'get it right' in attaining such as balance, as some of our findings reveal, with a tendency for new knowledge to be valued over traditional knowledge.

The potential for expanded use and adaptation of this curriculum as a teaching tool for building solidarity and working towards food sovereignty is itself dependent on structural issues such as financial support for the institutions using it, and the broader political and economic context in which these institutions are operating, including relations with the state (Meek et al. 2017). Africa is the site of intense efforts by corporations, foundations and research institutions to foment a 'new Green Revolution' (Moseley et al. 2015) and as such is a particularly challenging place to support radical pedagogical initiatives to foster food sovereignty using agroecology. Both Malawi and Tanzania have endorsed the Green Revolution approach through policies such as fertilizer and hybrid seed subsidies, and setting aside land for promoting large-scale agricultural intensification. At the same time initiatives by organizations such as the African Food Sovereignty Alliance and others provide alternative models, albeit with limited financial and political support. Given this context, it remains to be seen whether this curriculum can be truly used for transformative change towards food sovereignty.

Conclusion

In summary, this curriculum is novel for its use of interdisciplinary, cross-cultural and participatory methods to develop a relevant curriculum that can be taught *by* farmers *to*

Table 3 Key concepts and application in curriculum

Concepts	Summary of key ideas	Application in curriculum experience	Relevant references
Participatory action research	Successful collaboration requires that people treat each other as equals and as friends, respect each other's ideas and build on their experiences. Listening to each other is essential. None of us have all of the knowledge, that's why we need each other	Multiple opportunities for dialogue and exchange, including field visits, face-to-face meetings and exchanges of materials	(Freire and Ramos 1970; Werner and Bower 1982)
Social learning	Learning is multi-faceted and multi-directional, and participatory methods and reflection are important to include. Teaching methods should encourage co-learning, ability to move out of one's own comfort zone to learn from others, including divergent perspectives	Integration of different subjects in a given day; use of drama to encourage dialogue	(Cundill et al. 2014)
Theatre for development, participatory drama	Increased dialogue and social interaction, collaborative problem solving, increased comprehension and engagement, increased self and collective confidence around new knowledge and innovations	Pre-written stories to perform, and opportunity to create drama based on material	(Abah 1996; Bezner Kerr 1998; Sloman 2012)
Experiential learning	Hand-on activities such as observing what you see in the forest versus cultivated fields; testing different soil types, looking at insects	Hands-on activities and field visits to do observations	(Østergaard et al. 2010; Moncure and Francis 2011)
Transdisciplinary approaches	Seeing the connections between activities in the community and on the farm. Valuing the connection between activities may be more important as introducing new activities	Integrating different topics within the curriculum (several in one day); use of drama and stories to demonstrate how the topics are linked	(Méndez et al. 2013)
Feminist intersectional praxis	There are different forms of oppression that can intersect and affect people's available resources, time and perspective; Marginalized and oppressed groups have knowledge and experience that can help to solve their own problems, shape knowledge and understanding	Dialogue-based, problem-solving methodologies, attention to inequalities at multiple scales and the intersection of different forms of inequalities	(Naples 2013)
Food sovereignty	Autonomy; local knowledge valued; gender equity important; agroecological methods as praxis	Integration of local knowledge, attention to gender equity, agroecology, experiential learning, involvement of farmers in building curriculum	(Meek 2015; McCune et al. 2016; Patel 2012; Wittman 2010)

farmers in southern and eastern Africa. Preliminary results indicate that the resulting curriculum will help advance agroecology among smallholder farmers in low-resource settings, while simultaneously attending to, and improving gender equity and human and soil nutrition. This approach should benefit scholars, students and practitioners, as the basis for enhanced relevance, improved understanding of practical implementation of agroecology theory as it relates to food sovereignty (Østergaard et al. 2010; Meek et al. 2017). Ongoing efforts to sharpen critical pedagogical tools will be needed to effectively challenge dominant hegemonic narratives of modern agricultural systems in the region.

Acknowledgements We gratefully acknowledge the Atkinson Center for a Sustainable Future at Cornell University for funding this unorthodox project. Farmer representatives from the Soils, Food and Healthy Communities organization and the Malawi Farmer to Farmer Agroecology project provided key input into the ideas that led to this article. The authors of curriculum include several who did not participate in this article: Laurie Drinkwater, Shupo Kumwenda, Joanne Thiessen Martens and Elias Mtinda, and some technical input from Vernon Kabambe, Elisa Mazuma, Ambonisegwe Mbwaga and Kelvin Mtei. We also acknowledge the trainers in Malawi and Tanzania who piloted the curriculum: Anita Chitaya, Tinkani Gondwe, Esther Kalonga, Esther Maona, Malumbo Mithi, Mwapi Mkandawire, Rodgers Msachi, Blessings Nyirenda, Zacharia Nkhonya, Pressings Moyo, Innocent Mhoni, Burton Gama,Tanazio Moses, Seliya Jabesi, Alice Gubudu,Lesita Malisawo, Maliseni Kenneth, Kennedy Salimbira, Edwin Kasamba Nyathi ,Christina Hara and Paul Nkhonjera. There are many Cornell University students who contributed to the curriculum: Eleanore Baughan, Ann Lei, Valerie Ota (sadly deceased), Julia Schaffer, Tessa Schneider, Evlyn Samuel, and Sarah Zipfel.

References

Abusabha, Rayane, Jane Peacock, and Cheryl Achterberg. 1999. How to make nutrition education more meaningful through facilitated group discussions. *Journal of the American Dietetic Association* 99: 72–76. https://doi.org/10.1016/S0002-8223(99)00019-X.

Affleck, William, and Gretel, Pelto. 2012. Caregivers' responses to an intervention to improve young child feeding behaviors in rural Bangladesh: A mixed method study of the facilitators and barriers to change. *Social Science & Medicine* 75: 651–658. https://doi.org/10.1016/j.socscimed.2012.03.030.

Altieri, Miguel A. 1995. *Agroecology: the science of sustainable agriculture*. 2 ed. Boulder: Westview Press.

Arimond, Mary, Corinna Hawkes, Marie T. Ruel, Z. Sifri, R. Peter, J. L. Berti, Jan W. Leroy, L. R. Low, Brown, and Edward A. Frongillo. 2010. Agricultural interventions and nutrition: lessons from the past and new evidence. In *Combating micronutrient deficiencies*, eds. Brian Thompson, and Leslie Amoroso, 41–75. Rome: Food and Agricultural Organization (FAO).

Berti, P.R., Julia, Krasevec, and Sian, FitzGerald. 2004. A review of the effectiveness of agriculture interventions in improving nutrition outcomes. *Public Health Nutrition*. https://doi.org/10.1079/PHN2003595.

Bezner Kerr, Rachel. 2005. Food security in northern Malawi: Historical context and the significance of gender, kinship relations and entitlements. *Journal of Southern African Studies* 31: 53–74.

Bezner Kerr, Rachel. 2014. Lost and found crops: Agrobiodiversity, indigenous knowledge, and a feminist political ecology of sorghum and finger millet in northern Malawi. *Annals of the Association of American Geographers* 104 (3): 577–593. https://doi.org/10.1080/00045608.2014.892346.

Bezner Kerr, Rachel. 2017. Gender and agrarian inequalities: From global systems to household relations. In *Agricultural systems: agroecology and rural development*, eds. Sieglinde S. Snapp, and Barry Pound, 2nd ed., 333–370. Boston: Elsevier.

Bezner Kerr, Rachel, Sieglinde, Snapp, Lizzie, Shumba, Marko Chirwa, Lizzie Shumba and Rodgers Msachi. 2007. Participatory research on legume diversification with Malawian smallholder farmers for improved human nutrition and soil fertility. *Experimental Agriculture* 43 (4): 1–17. https://doi.org/10.1017/S0014479707005339.

Bezner Kerr, Rachel, Laifolo, Dakishoni and Lizzie, Shumba. 2008. We grandmothers know plenty': Breastfeeding, complementary feeding and the multifaceted role of grandmothers in Malawi. *Social Science and Medicine* 66 (5): 1095–1105.

Bezner Kerr, Rachel, Peter, R., Berti, and Lizzie Shumba. 2011. Effects of a participatory agriculture and nutrition education project on child growth in northern Malawi. *Public Health Nutrition* 14: 1466–1472. https://doi.org/10.1017/S1368980010002545.

Bezner Kerr, Rachel, Rodgers, Msachi, Laifolo Dakishoni, Lizzie Shumba, Zachariah Nkhonya, Peter R. Berti and Christine Bonatsos et al. 2012. Growing healthy communities: Farmer participatory research to improve child nutrition, food security and soils in Ekwendeni, Malawi. In *Ecohealth research in practice: Innovative applications of an ecosystem approach to health*, ed. Dominique F. Charron, 37–46. Ottawa/New York: IDRC/Springer.

Bezner Kerr, Rachel, Lizzie, Shumba, Laifolo, Dakishoni, Esther, Lupafya, Peter R. Berti and Sieglinde S., Snapp, et al. 2014. Participatory, agroecological and gender-sensitive approaches to improved nutrition: A case study in Malawi. Invited submission to the FAO Expert Meeting in November 2013, 'Nutrition-Sensitive Food and Agriculture Systems' in preparation for ICN + 21. http://www.fao.org/food/nutritional-policies-strategies/icn2/expert-papers.

Bezner Kerr, Rachel, Hanson, Nyantakyi-Frimpong, Esther, Lupafya, Laifolo, Dakishoni, Lizzie, Shumba, Isaac, Luginaah. 2016. Building resilience in African smallholder farming communities through farmer-led agroecological methods. In *Climate change and agricultural development: Improving resilience through climate smart agriculture, agroecology and conservation*, ed. Udaya Sekhar Nagothu, 109–130. London: Routledge.

Bezner Kerr, Rachel, Hanson, Nyantakyi-Frimpong, Esther, Lupafya, Lizzie, Shumba, Isaac, Luginaah, Laifolo, Dakishoni and Sieglinde S., Snapp. 2018. Knowledge politics in participatory climate change adaptation research on agroecology in Malawi. *Renewal Agriculture and Food Systems*, 33: 238-251.

Chinsinga, Blessings. 2011. Seeds and subsidies: The political economy of input programmes in Malawi. *IDS Bulletin* 42 (4): 59–68. https://doi.org/10.1111/j.1759-5436.2011.00236.x.

Chirwa, Ephraim W., & Andrew Dorward. 2013. *Agricultural input subsidies: the recent Malawi experience*. 1st ed. Oxford: Oxford University Press.

Chowa, Clodina, Chris Garforth, and Sarah Cardey. 2013. Farmer experience of pluralistic agricultural extension, Malawi. *Journal of Agricultural Education and Extension* 19 (2): 147–166.

Collins, Patricia Hill. 1991. Black feminist thought: Knowledge, consciousness, and the politics of empowerment. In *Perspectives on gender 2*. New York, NY: Routledge.

De Schutter, Olivier. 2013. *United Nations Pecial Rapporteur for the Right to Food: Mission to Malawi from 12 to 22 July 2013-End of Mission Statement*. Geneva, Switzerland: United Nations.

Drinkwater, Laurie E., and S. Sieglinde, Snapp. 2008.. Nutrients in agroecosystems: Rethinking the management paradigm. *Advances in Agronomy* 92: 163–186.

Ellis, Frank, and Elizabeth Manda. 2012. Seasonal food crises and policy responses: A narrative account of three food security crises in Malawi. *World Development* 40: 1407–1417. https://doi.org/10.1016/j.worlddev.2012.03.005.

Fals-Borda, Orlando, and Muhammad Anisur Rahman, eds. 1991. *Action and knowledge: breaking the monopoly with participatory action- research*. New York: Apex Press.

Freire, Paulo. 1970. *Pedagogy of the oppressed. The political economy of development and underdevelopment*: Translator: Myra Bergman Ramos, 557–575.

Funk, Chris, Michael D Dettinger, Joel C Michaelsen, James P Verdin, Molly E Brown, and Matthew Barlow, and Andrew Hoell. 2008. Warming of the Indian Ocean threatens eastern and southern African food security but could be mitigated by agricultural development. *Proceedings of the National Academy of Sciences of the United States of America* 105: 11081–11086.

Harding, Sandra G. 1986. *The science question in feminism*. Ithaca: Cornell University Press.

Hartsock, Nancy. 1999. Postmodernism and political change: Issues for feminist theory. Cultural Critique: 15. https://doi.org/10.2307/1354291.

Hassim, Shireen. 2009. After apartheid: Consensus, contention, and gender in South Africa's public sphere. *International Journal of Politics, Culture, and Society*. https://doi.org/10.1007/s10767-009-9076-6.

Kangalawe, Richard Y.M., Carl Christiansson, and Östberg Wilhelm Wilhelm. 2008. Changing land-use patterns and farming strategies in the degraded environment of the Irangi Hills, central Tanzania. *Agriculture, Ecosystems & Environment* 125: 33–47. https://doi.org/10.1016/j.agee.2007.10.008.

Kerr, David. 1995. *African popular theatre: From postcolonial to the present*. London: James Currey.

Knueppel, Danielle, Montague Demment, and Lucia Kaiser. 2010. Validation of the household food insecurity access scale in rural Tanzania. *Public Health Nutrition* 13: 360. https://doi.org/10.1017/S1368980009991121.

Lin, Brenda B. 2011. Resilience in agriculture through crop diversification: Adaptive management for environmental change. *BioScience* 61: 183–193. https://doi.org/10.1525/bio.2011.61.3.4.

Malawi. Ministry of Agriculture, Irrigation and Water Development. (MAIWD). 2015. *Nutrition handbook for farmer field schools*. Lilongwe: GPO. Print.

McCune, Nils, Juan Reardon, and Peter Rosset. 2014. Agroecological formación in rural social movements. *Radical Teacher* 98: 31–37. https://doi.org/10.5195/RT.2014.71.

McCune, Nils, Peter M. Rosset, Helda Morales, Tania Cruz Salazar and Antonio Saldívar Moreno. 2016. Mediated territoriality: rural workers and the efforts to scale out agroecology in Nicaragua. *The Journal of Peasant Studies*. https://doi.org/10.1080/03066150.2016.1233868.

McCune, Nils, Peter M. Rosset, Tania Cruz Salazar, Helda Morales, and Moreno Antonio Saldívar. 2017. The long road: Rural youth, farming and agroecological *formación* in Central America. *Mind, Culture, and Activity*. https://doi.org/10.1080/10749039.2017.1293690.

Meek, David. 2015. Learning as territoriality: the political ecology of education in the Brazilian landless workers' movement. *The Journal of Peasant Studies* 42: 1179–1200. https://doi.org/10.1080/03066150.2014.978299.

Meek, David, Peter, Rebecca Tarlau, Katharine, Bradley, Bruce, Ferguson, Lesli, Hoey, Helda, Morales, P., Rosset. 2017. Food sovereignty education across the Americas: Multiple origins, converging movements. *Agriculture and Human Values*. https://doi.org/10.1007/s10460-017-9780-1.

Méndez, V., Ernesto, Christopher M, Bacon, and Roseann Cohen. 2013. Agroecology as a transdisciplinary, participatory, and action-oriented approach. *Agroecology & Sustainable Food Systems* 37: 3–18.

Menza, Valeria and Claudia Probart. 2013. *Eating well for good health: Lessons on nutrition and healthy diets*. Rome: FAO.

Messina, J. P., B. Peter, S. Sieglinde Snapp. 2017. The myth of the Malawian farm input subsidy program. *Nature Plants* 3 (4): 17013. https://doi.org/10.1038/nplants.2017.13.

Mistry, Jayalaxshmi, and Andrea Berardi. 2016. Bridging indigenous and scientific knowledge. *Science* 352: 1274–1275. https://doi.org/10.1126/science.aaf1160.

Moncure, Shannon, and Charles Francis. 2011. Foundations of experiential education as applied to agroecology. *NACTA Journal* 55: 75–91.

Morrone, Vicki. 2017. Outreach for rural innovation. Agricultural systems: Agroecology and rural innovation for development. eds. Snapp, Sieglinde S. and Barry Pound. 2 ed., 407–439, Academic Press: Cambridge.

Moseley, William, Matthew Schnurr, and Rachel Bezner Kerr. 2015. Interrogating the technocratic (neoliberal) agenda for agricultural development and hunger alleviation in Africa. *African Geographical Review*, 1–7. https://doi.org/10.1080/19376812.2014.1003308.

Moyo, Boyson Henry Zondiwe and Dumisani Zondiwe Moyo. 2014. Indigenous knowledge perceptions and development practice in Northern Malawi: Indigenous knowledge and development. *The Geographical Journal* 180 (4): 392–401.

Msachi, Rodgers, Laifolo Dakishoni and Rachel Bezner Kerr. 2009. Soils, food and healthy communities: working towards food sovereignty in Malawi. *Journal of Peasant Studies* 36 (3): 700–706.

Mulwafu, Wapulumuka Oliver. 2015. Conservation song: A history of peasant-state relations and the environment in Malawi, 1860–2000. *Environmental History* 20 (2): 205–207. https://doi.org/10.1093/envhis/emv012.

Naples, Nancy A. 2013. Sustaining democracy: Localization, globalization, and feminist praxis. *Sociological Forum* 28: 657–681. https://doi.org/10.1111/socf.12054.

National Statistical Office (NSO) [Malawi] and ICF. 2017. Malawi demographic and health survey 2015-16. Zomba, Malawi, and Rockville, Maryland, USA. NSO and ICF.

Nyantakyi-Frimpong, Hanson, Catherine Hickey, Esther Lupafya, Laifolo Dakishoni, Rachel Bezner Kerr, Blessing Nyirenda, Zachariah Nkhonya, Mangani Katundu and George Gondwe. 2017. A farmer-to-farmer agroecological approach to addressing food security in Malawi. In *Everyday Experts: How people's knowledge can transform the food system*, eds. Tom Wakeford, Javier Sanchez-Rodriguez, Marina, Chang Christabel Buchanan and Colin Anderson, 119–134. Coventry: Center for Agroecology, Water and Resilience, Coventry University.

Okoth, James Robert, and Winfred Nalyongo. 2013. *Facilitators' guide for running an agro pastoral field school: An adaptation to agro-pastoral setting*. Uganda: FAO.

Østergaard, Edvin, Geir Lieblein, Tor Arvid Breland, and Charles Francis. 2010. Students learning qgroecology: Phenomenon-based education for responsible action. *The Journal of Agricultural Education and Extension* 16: 23–37. https://doi.org/10.1080/13892240903533053.

Patel, Raj, Rachel, Bezner Kerr, Lizzie Shumba, and Laifolo Dakishoni. 2015. Cook, eat, man, woman: understanding the New Alliance for Food Security and Nutrition, nutritionism and its alternatives from Malawi. *The Journal of Peasant Studies* 42: 21–44. https://doi.org/10.1080/03066150.2014.971767.

PATH and CARE. 2011. *Infant and young child feeding and gender: A training manual for male group leaders*. Washington, DC: PATH and CARE.

Peterman, Amber. 2011. Women's property rights and gendered policies: Implications for women's long-term welfare in rural

Tanzania. *Journal of Development Studies* 47: 1–30. https://doi.org/10.1080/00220381003600366.

Popkin, Gabriel. 2016. Collaborations: Partners in knowledge. *Nature* 535: 581–582.

Qureshi, M., Elaz, John, Dixon and Mellissa, Wood. 2015. Public policies for improving food and nutrition security at different scales. *Food Security*, 7(2), 393–403. https://doi.org/10.1007/s12571-015-0443-z.

Reynolds, Kristin, and Nevin, Cohen. 2016. *Beyond the kale: Urban agriculture and social justice activism in New York City*. Athens: The University of Georgia Press.

Rosset, Peter Michael, Braulio, Machín Sosa, Adilén, María, Roque Jaime and Dana, Rocío, Ávila Lozano. 2011. The Campesino-to-Campesino agroecology movement of ANAP in Cuba: Social process methodology in the construction of sustainable peasant agriculture and food sovereignty. *The Journal of Peasant Studies* 38:161–191.

RWANMREC and PROMUNDO. 2014. *Engaging men as fathers in gender equality, maternal and child health, caregiving and violence prevention*. Washington D.C.: Rwanda Men's Resource Center, Kigali, Rwanda and Promundo-US.

Sloman, Annie. 2012. Using participatory theatre in international community development. *Community Development Journal* 47: 42–57. https://doi.org/10.1093/cdj/bsq059.

Snapp, Sieglinde, and K. L. Heong. 2003. Scaling up: participatory research and extension to reach more farmers. In *Uniting science and participation: Managing natural resources for sustainable livelihoods*, eds. Barry, Pound, Sieglinde, Snapp, Cynthia, McDougall and Ann, Braun, 67–87. UK: Earthscan and Canada: IRDC.

Snapp, Sieglinde S., J. Malcolm, Blackie, Robert A., Gilbert, Rachel, Bezner Kerr and Y. George Kanyama-Phiri. 2010. Biodiversity can support a greener revolution in Africa. *Proceedings of the National Academy of Sciences* 107: 20840–20845. https://doi.org/10.1073/pnas.1007199107.

SUSTAINET, E. A. 2010. Technical manual for farmers and field extension service providers: Farmer field school approach. Sustainable Agriculture Information Initiative, Nairobi, Kenya. http://www.sustainetea.org.

Tschakert, Petra, and Kathleen Dietrich. 2010. Anticipatory learning for climate change adaptation and resilience. *Ecology and society* 15:11.

UNICEF. 2013. *The community infant and young child feeding counselling package*. New York: UNICEF.

Werner, David and Bill, Bower. 1991. *Helping health workers learn: A book of methods, aids, and ideas for instructors at the village level*. 9th ed. Palo Alto: The Hesperian Foundation.

Wolfe, David W. 2013. Climate change solutions from the agronomy perspective. In Handbook of climate change and agroecosystems: Global and regional aspects and implications, Daniel Hillel, and Cynthia Rosenzweig, eds. ICP Series on Climate Change Impacts, Adaptation, and Mitigation vol. 2. Hackensack, New Jersey: World Scientific.

Zulu, Leo Charles. 2010. The forbidden fuel: Charcoal, urban woodfuel demand and supply dynamics, community forest management and woodfuel policy in Malawi. *Energy Policy* 38 (7): 3717–3730.

Publisher's Note Springer Nature remains neutral with regard to jurisdictional claims in published maps and institutional affiliations.

Rachel Bezner Kerr is an Associate Professor in Development Sociology at Cornell University. She currently serves on a project team of the High Level Panel of Experts on Food Security & Nutrition of the United Nations and the Coordinating Lead Author for Chapter 5 on Food for the next Intergovernmental Panel on Climate Change report. She does research using participatory methodologies to test the impacts of agroecological approaches on livelihoods, nutrition and sustainable land management for rural communities in Malawi and Tanzania.

Sera L. Young is an assistant professor of anthropology and global health at Northwestern University. Methodologically, she draws on her training in medical anthropology, international nutrition, and HIV to take a biocultural approach to understanding how mothers in low- and middle-income countries cope to preserve their health and that of their families (www.serayoung.org). Currently, she is studying barriers to optimal infant and young child feeding, including both food and water insecurity. She leads the Household Water InSecurity Experiences (HWISE) research consortium, a network that seeks to develop a novel scale to comparably measure household-level water insecurity across diverse ecological settings. It is anticipated that this scale, like the development of a household-level food insecurity metric, will be transformative for both academia and policy.

Carrie Young received her PhD in Communication (Cornell '17) with a focus on food security, climate change, and other related sustainability issues from a social and behavioral perspective. She served in the Peace Corps in Mali, West Africa, where she lived in a small subsistence farming village, and saw firsthand many of the urgent issues facing farmers. She also worked as a researcher for National Geographic, and extensive time spent on the magazine's climate change issue solidified the value of projects like the farmer-engaged integrated curriculum for her.

Marianne V. Santoso is a PhD Candidate in International Nutrition at Cornell University's Division of Nutritional Sciences and expects to graduate in May 2019. Her dissertation research explores the role of intra-household gender equity, especially on household decision-making and task division, in ensuring optimum resource (both time and economic) allocation for child nutrition. She also manages Singida Nutrition and Agroecology Project (SNAP-Tz), a participatory research project engaging farmers in Singida, Tanzania in sustainable agriculture practices (agroecology), nutrition, and gender roles.

Mufunanji Magalasi is an Associate Professor of Drama and Development Media at the University of Malawi. He is currently a Co-Case Country Manager for Malawi in the EU Horizon 2020 project, InnovAfrica, and a collaborating Communication Specialist for an IDRC Fish Value Chain Project run at the University of Malawi.

Martin Entz is Professor of Cropping Systems and Natural Systems Agriculture in the Faculty of Agricultural and Food Sciences at the University of Manitoba. He is founder and scientific lead of the Glenlea long-term plots, Canada's oldest organic field experiment. Martin does research using participatory approaches and the work includes a Canadian farmer-participatory plant breeding program and a cropping system co-design program. He provides science support to the Canadian Foodgrains Bank, including development agronomy research in Eastern and Southern Africa.

Esther Lupafya is a Health and Gender Coordinator at Soils Food and Health Communities (SFHC) Organization. She is one of the facilitators for Curriculum training using drama in the community. Esther Lupafya is a community nurse and holds an M.A. in Social Development and Health obtained from Queen Margaret University (Scotland).

Laifolo Dakishoni is the Research Coordinator of the non-profit Soils, Food and Healthy Communities (SFHC) organization in Malawi. SFHC carries out participatory research with smallholder farmers on agroecological and community-based approaches to address food security, nutrition and social equity. He has co-authored several articles on this work.

Vicki Morrone has been with Michigan State University since 2002. She is a member of the Center for Regional Food Systems and focuses on organic production systems and on-farm research assisting farmers to address the needs of their soil and meet organic system needs. She is passionate about helping farmers visualize their needs through on-farm assessments and see new possibilities. Her work takes her to Michigan and Eastern Africa. Morrone has a Master of Science in Plant Pathology and Extension Services from Penn State University.

David Wolfe is Professor in the School of Integrative Plant Science at Cornell University. His research and outreach efforts focus primarily on agricultural climate change adaptation and mitigation, and soil health and water management. In addition to peer-reviewed research publications, science communication efforts include analyses for policy-makers, such as co-authoring the 2008 and 2014 National Climate Assessments, and writing commentary for the popular press. At Cornell he teaches courses on climate change and environmental science writing, is lead project director for the New York Soil Health program, and is on the Advisory Board for the Cornell Institute for Climate Smart Solutions.

Sieglinde Snapp is a Professor of Soils and Cropping Systems Ecology and Associate Director of the Center for Global Change, Michigan State University. She has been engaged in participatory action research and extension since 1993 in Malawi, in support of co-generation of agroecological knowledge and diversified, sustainable food production systems [http://globalchangescience.org/eastafricanode]. Learning with farmers, being a grandmother and co-editing the text book 'Agricultural Systems' are among her greatest privileges.

Affiliations

Rachel Bezner Kerr[1] · Sera L. Young[2] · Carrie Young[3] · Marianne V. Santoso[4] · Mufunanji Magalasi[5] · Martin Entz[6] · Esther Lupafya[7] · Laifolo Dakishoni[7] · Vicki Morrone[8] · David Wolfe[9] · Sieglinde S. Snapp[10]

Sera L. Young
sera.young@northwestern.edu

Carrie Young
cey26@cornell.edu

Marianne V. Santoso
mvs43@cornell.edu

Mufunanji Magalasi
mufunanjimagalasi@gmail.com

Martin Entz
M_Entz@umanitoba.ca

Esther Lupafya
elupafya@gmail.com

Laifolo Dakishoni
dakishoni@gmail.com

Vicki Morrone
sorrone@anr.msu.edu

David Wolfe
dww5@cornell.edu

Sieglinde S. Snapp
snapp@msu.edu

1 Department of Development Sociology, Cornell University, 262 Warren Hall, Ithaca, NY 14853, USA

2 Department of Anthropology, Northwestern University, 515 Clark Street, Evanston, IL 60208, USA

3 Department of Communication, Cornell University, 498 Mann Library, Ithaca, NY 1485, USA

4 Division of Nutritional Sciences, Cornell University, S-2072, Schurman Hall, Ithaca, NY 14853, USA

5 Department of Fine and Performing Arts, University of Malawi, P.O Box 280, Zomba, Malawi

6 Department of Plant Science, University of Manitoba, Winnipeg, MB R3T 2N2, Canada

7 Soils, Food and Healthy Communities, PO Box 36, Ekwendeni, Malawi

8 Community Sustainability, Michigan State University, 480 Wilson Road, East Lansing, MI 48824, USA

9 School of Integrative Plant Sciences, Cornell University, 135 Plant Science Building, Ithaca, NY 14853, USA

10 Department of Plant, Soil and Microbial Sciences, Michigan State University, 1066 Bogue Street, East Lansing, MI 48824, USA

Agriculture and Human Values (2019) 36:567–579
https://doi.org/10.1007/s10460-018-9863-7

SYMPOSIUM/SPECIAL ISSUE

Multi-actor networks and innovation niches: university training for local Agroecological Dynamization

Daniel López-García[1] · Laura Calvet-Mir[2,3] · Marina Di Masso[4] · Josep Espluga[5]

Accepted: 10 April 2018 / Published online: 2 May 2018

Abstract

The global environmental and social-economic crises of industrialized agriculture have led to the emergence of agroecology as an alternative approach aiming to increase the ecological, social and economic sustainability of agri–food systems. The 'multi-level perspective' is now a widely used framework to understand and promote the upscaling of local innovation niches, such as agroecology, to broader scales (e.g., regional, national, international), thus reconfiguring the dominant socio-technical regimes. Additionally, emergent 'hybrid forums' can provide a space between niche and regime where niche innovators can become important actors in scaling up and out emergent innovations. In this paper, we examine a university training program (Postgraduate Diploma in Local Agroecological Dynamization at the Universitat Autònoma de Barcelona), to better understand its role as a 'hybrid forum'. Our analysis focuses especially on how the program, as an example of a hybrid forum, worked to reconfigure practices, concepts, and tools of local development practitioners. We also assess to what extent the program contributed to transitioning local development institutions toward agroecology. An online survey (n = 46) and in-depth interviews (n = 16) were carried out to determine how the training program has impacted the student's opinions and their respective institutions. The results show that most of the students consider that they have acquired new theoretical frameworks and useful methods to re-framing their local development projects, that new alliances with multi-actor networks have been perceived, and that some internal changes of the local development practices have taken place. We conclude that the training program, as a hybrid forum, is capable of outscaling niche innovations through linkages with different kind of actors both from the niche and the regime. Political changes in the socio-technical landscape level offer an opportunity to amplify the impact of the innovations which are being generated by those multi-actor networks, but with a limited multi-level impact as far as institutional regime-actors not aligned with agroecological transition keep the most of the competencies on agri–food systems.

Keywords Agroecology · Agroecological transitions · Hybrid actors · Hybrid forums · Multi-level perspective · Grassroots innovation and social innovation

Abbreviations

LAeD	Local Agroecological Dynamization
MLP	Multi-level perspective
NGO	Non-governmental organization

✉ Daniel López-García
daniel@entretantos.org

Laura Calvet-Mir
lcalvetmir@gmail.com

Marina Di Masso
madimasso@gmail.com

Josep Espluga
joseplluis.espluga@uab.cat

1 Fundación Entretantos, c/Arzobispo José Delicado 1, 47014 Valladolid, Spain

2 Internet Interdisciplinary Institute (IN3), Universitat Oberta de Catalunya, Barcelona, Spain

3 Institut de Ciència i Tecnologia Ambiental, Universitat Autònoma de Barcelona, Edifici ICTA-ICP, carrer de les Columnes s/n, 08193 Bellaterra, Barcelona, Spain

4 Agroecology and Alternative Food Systems Chair, Universitat de Vic-Universitat Central de Catalunya, Carrer de la Sagrada Família, 7, Vic, 08500 Catalonia, Spain

5 Facultat de CCPP i Sociologia, Institut de Govern i Polítiques Públiques, Universitat Autònoma de Barcelona, Edifici B, Campus de la UAB, Bellaterra, 08193 Catalonia, Spain

Introduction

Local Agroecological Dynamization to promote agroecological transitions

The worldwide environmental and social–economic crisis of industrialized agriculture has led to the emergence of agroecology as an alternative approach aiming to increase agri–food systems' sustainability from an ecological, social and economic perspective (Francis et al. 2003). Agroecology was early defined as a theoretical and methodological framework to apply ecological concepts and principles to the design of sustainable agricultural systems (Altieri 1983; Gliessman 2002). Later developments addressed the need to also integrate the cultural, economic and social features of agroecosystems, as farming sustainability challenges are deeply conditioned by other non-ecological processes (Guzmán et al. 2000). According to Méndez et al. (2016), there still are two predominant perspectives on agroecology: one focused on agronomic and ecological processes related to farming systems, and another focused on wider agri–food system issues which engage natural sciences with social sciences. In this paper, we follow the second perspective as it takes into account the different features affecting agri–food systems (i.e., ecological, socio-cultural and political), and thus addresses a more comprehensive approach to its sustainability.

As an action-oriented approach, agroecology seeks scaling-out agri–food systems' sustainability in a process called 'agroecological transition' (Méndez et al. 2016). Several Spanish case studies stress that challenges for agroecological transition are often not only ecological but related to socio-cultural and political dimensions of agroecology (Guzmán et al. 2016). Indeed, the transition from industrialized to agroecological systems is not smooth (Lobley et al. 2009; Milestad et al. 2010). It requires specific extension practices adapted to entirely different systems of agricultural management, which also needs a collective process of individual and social learning (Röling and Wagemakers 1998; Cuéllar-Padilla and Calle-Collado 2011). Thus, the classical innovations adoption model used for agricultural modernization (Rogers 1962) is not readily adapted to agroecology (Padel 2001, 2008; Heleba et al. 2016). A more complex approach is required which articulates the ecological-productive dimension of agroecology (farm scale) with other dimensions corresponding to higher scales (i.e., rural development and socio-political dimensions) (Guzmán et al. 2016).

Local Agroecological Dynamization (LAeD) is a strategy for promoting local sovereignty through a set of methods and tools that address the resilience and empowerment of local communities, and incorporates a participative and territorial approach. The LAeD was proposed in the second decade of the 2000s as a practical tool to promote agri–food systems' sustainability at a local scale through agroecological transition processes (see Lopez-Garcia and Guzman Casado 2014; López-García et al. 2015 for an in-depth knowledge of its origins). LAeD mobilizes local communities' actor-networks, resources, and capabilities through the reactivation of local agricultural production. LAeD also promotes leadership amongst farmers, multi-actor networking, agroecological education, traditional agroecological knowledge, conservation and alternative food networks. For this purpose, it articulates participatory action-research methodologies with other research and community development methods to enhance local communities' capabilities to build up appropriate, comprehensive, inclusive and territory-based agroecological projects.

Agroecological transitions from a multi-level perspective

Within a holistic approach, agroecological transitions might be understood as comprehensive processes of agri–food system socio-technical transition. Thus, as a radical (but progressive) shift from one socio-technical regime to another, involving a change in the technology, infrastructures, cultures, rules, values, institutions and practices of a societal system (Geels 2002, 2010; Darnhofer 2015). In recent years, a growing amount of literature on agri–food transitions to sustainability was produced following the multi-level perspective (MLP). MLP has been noted as an appropriate strategy to understand and promote the upscaling of local agroecological transition processes to regional or higher administration levels (Levidow et al. 2014). Such an approach covers two main gaps on agroecological transitions research (Sanderson and Ioris 2017): (1) an excessive focus on the farm level, in which agroecological transition is sometimes confused with conversion to organic farming, and regime–niche interactions are not explicitly addressed (see Guzmán et al. 2000, 2013; Gliessman and Rosenmeyer 2010). (2) The development of methodological tools without a robust theoretical framework on socio-technical transitions (see Wezel et al. 2015; Duru et al. 2015; Méndez et al. 2016).

From the MLP, transitions to sustainability are 'non-linear processes that result from the interplay of developments at three analytical levels: niches (the locus for radical innovations), socio-technical regimes (the locus of established practices and associated rules that stabilize existing systems), and exogenous (socio-technical) landscapes' (Geels 2010, p. 26). Within agroecology, niches are social spaces where multi-actor alliances (e.g., farmers, consumers, NGOs and potentially extension or research services) protect farming innovations linked to alternative food networks and new institutions for agri–agri–food system sustainability (Levidow et al. 2014; Méndez et al. 2016). The ability of such networks to address the social, cultural, economic and ecological challenges of agri–food systems within a comprehensive framework is what creates the transformative character of innovations. It thus leads to stable regime reconfigurations (Levidow et al. 2014) accomplished through a progressive, diverse and complex process of niche–regime interactions which include practices and actors (Geels 2010, 2011; Darnhofer 2015).

Regimes are neither monolithic nor free of internal conflicts, but they change following a co-evolutionary pattern in interaction with niches and socio-technical landscapes (Geels 2011; Diaz et al. 2013; Darnhofer 2015). Within time, regimes can show performance weaknesses which introduce pressure for innovation. Once a window of opportunity appears out of regime malfunctioning, innovation niches emerge which can lead to regime reconfigurations (Elzen et al. 2011; Diaz et al. 2013). Each of these above mentioned analytical levels should be aligned to allow niche-innovations to break-through into regime reconfiguration (Darnhofer 2015). However, the interrelations between niche and regimes which lead to effective, radical transitions are still under discussion. Elzen et al. (2012) distinguish the process of 'anchoring' when a novelty spreads and becomes an innovation niche and 'linking' when a niche establishes durable links with a regime. They propose three types of anchoring: (1) *technological*, when a technological novelty is introduced in a productive sector or territory, (2) *network*, when the novelty spreads over additional territories or sectors through actors' intensified contact and knowledge exchange, and (3) *institutional*, when the diffusion of the novelty is adopted by institutions and somehow supported by public policies, even if only in a marginal way. Bui et al. (2016) establish three stages for regime reconfiguration: (1) the emergence of alternative initiatives, (2) the construction of a socio-technical niche through the enrollment of new actors, and (3) the construction of an alternative model, which impacts public bodies and generates related public policies.

An overall consensus can be observed regarding the collaborative nature of innovation processes on agri–food sustainability transitions and on the central role of multi-actor networks beyond the role of individuals (Ploeg and Marsden 2008; Lamine 2011; Neumeier 2012; Bui et al. 2016). Multi-actor networks' role in transitioning to sustainability is twofold: (1) as an operational core element on the maturation of novelties ('network anchoring') and the upscaling of niches into regime reconfiguration (Tisenkopfs et al. 2009; Elzen et al. 2012; Bui et al. 2016), and (2) as a conduit to strengthen the transformative elements of innovations and to create a protective environment for niche-actors to face regime-actors' pressures against regime reconfiguration (Diaz et al. 2013; Levidow et al. 2014; Darnhofer 2015). The 'innovation broker' profile would respond, therefore, to capabilities and skills oriented to the activation and reconfiguration of such networks and the associated cooperative processes (Galli and Brunori 2011). Such profiles could benefit by being 'hybrid actors', as their primary function is the creation of new alliances (Roep and Wiskerke 2004; Elzen et al. 2012; Diaz et al. 2013). Hybrid actors show characteristics of belonging to the regime as well as to the niche (Elzen et al. 2012). The challenge of niche-actors is thus identifying the hybrid actors and enrolling them in the socio-technical transition (Diaz et al. 2013) through the creation of a hybrid forum (Elzen et al. 2012). A hybrid forum is an area where niche and regime overlap, characterized by relatively stabilized innovation networks (resulting from network anchoring) where regime and niche developments come together at the most concrete level.

Our objective in this paper is to examine a university training program (Postgraduate Diploma in Local Agroecological Dynamization at the Universitat Autònoma de Barcelona), to better understand its role as a 'hybrid forum' which acts as a tool for promoting agroecological transitions, to accomplish the following: (i) reconfigure the hegemonic socio-technical regime, (ii) contribute to the reconfiguration of regime practices, (iii) expand agroecological concepts and tools for local development practitioners, and (iv) lead local development institutions towards an agroecological transition. In the following section, we will introduce the assessed agroecological university training, its aims and primary challenges. Section three briefly describes the case study and explains the methods used to obtain the results exposed in section four, while section five discusses these results within the framework of MLP. Finally, we conclude by tackling the impact of the university post-graduate training program in agroecological transitions, and discussing new insights on the application of the MLP to agroecological transitions framework.

Training for agroecological transitions through multi-actor networks

According to several authors (e.g., Guzmán et al. 2016; Padel 2001, 2008; Wheeler 2008), agroecological transitions need new approaches beyond the common agricultural extension practices. Tackling transition processes means sharing knowledge and enhancing capacities for adapting to change in a non-hierarchical or formalized manner (Woodhill 2009), as well as building capacities to design, lead, facilitate and support such processes in ways that lead to real learning and change. In this context, learning networks are key places to create heterogeneous groups of stakeholders which develop mutual trust and social cohesion (Vogelezang et al. 2009).

Following Heleba et al. (2016, p. 180), new agroecological extension practices stress the need of focusing on end-user needs, regarding design and implementation of learning processes; and to "meet the farmers where they are". Other authors stress the difference of adult-learners' needs from those of younger students, thus shaping educational processes focusing on prior experiences and motivations, and practical tools as case-studies or problem based learning (Ota et al. 2006; Franz 2007; Bell and McAllister 2012; cited in Heleba et al. 2016). In this sense, an agroecological training program based on students' real-world interactions could be an agroecological transition trigger (Francis et al. 2012).

Heleba et al. (2016) include the need for practitioners with skills in setting up "relationship architectures". Existing multi-actor networks in study sites are an additional variable which practitioners must address and assess. Experiencing complexity and interdisciplinary perspectives early in the process of defining the problem become an important element of phenomenological learning itself (Lieblein et al. 2012). In this sense, agroecological extension training programs can play the role of a 'hybrid forum' (Elzen et al. 2012) bringing together different localized agri–food system actors (i.e., local authorities, researchers, private actors and food movements). Within such hybrid forums, each actor's vision and interests are confronted with those of the other, serving as a representation of social diversity as it pertains to agri–food systems. Ultimately, this is intended to amplify the interactions among niche and regime-actors and discourses. Responses and arrangements to the complex challenges of agri–food systems need to be built upon a complex network of expertise, subjectivities, and positions. This network includes unequal positions of power and agency regarding socio-technical transitions. Within such forums, facilitated by researchers, innovative visions should lead to innovative proposals for transition paths.

The "end-user-needs focusing" approach uncovers a conflict related to the nature of local actors' expressed needs and practitioners' perceived needs. As demonstrated by Freire (1975, 2005), the expressed needs of the "oppressed" tend to show a strong "adherence" to oppressors' discourses and interests. Therefore, expressed needs of subaltern actors along the agri–food system often address conventional, close-minded solutions for its sustainability. Freire developed both a methodological (*Educaçao Popular*) and theoretical (*Pedagogia do oprimido*) approach to overcoming such adherence, which was also adapted to farming systems as an early criticism of rural extension (Freire 1969). This framework has been recognized as the core of some relevant methodological approaches for sustainable, participatory rural extension, namely Participatory Action-Research, Participatory Rural Appraisal, Farmers' Participatory Research, and the *Campesino a Campesino* Movement (Rhoades and Booth 1982; Bunch 1985; Farrington and Martin 1988; Chambers 1994a, b, c; Holt-Gimenez 2008; López Vargas et al. 2009; Guzmán et al. 2013; Méndez et al. 2016).

Within the current power structure of European agri–food systems, small local farmers can be considered as a kind of oppressed actors (Marsden and Sonnino 2008; Reed 2008; Bell et al. 2010). Despite small farmers being excluded from globalized agri–food networks, their adherence to oppressors' discourses and practices is observed both as a strong commitment to intensification discourses, and as a rejection of a shift towards more sustainable farming and added-value marketing strategies through alternative food networks (Darnhofer et al. 2005; Padel 2008; Guzmán et al. 2016; Meek 2015; Méndez et al. 2017). In such context, Freire's (2004) proposal to learn to "read the World" while "reading the word" should be at the core of a training program for agroecological extension to overcome such adherence to the globalized, conventional agri–food system. Following Freire's developments, the aim of participatory agroecological transition processes and methods should be, not only to ask the people but to create the inter-subjective environment in which people (local communities) develop their critical vision of their world. With this aim, agroecological practitioners should learn how to foster processes of action–reflection–action (Freire 2005) among small farmers and other local, rural actors to activate transitions to agri–food sustainability.

Methodology

Case study

Agroecology was firstly introduced in Spain through a PhD Program which begun in 1996, at Universidad Internacional de Andalucía. Such program tried to bring to Europe the novel developments of agroecology from the Americas, and to train a new generation of Latin-American activists, researchers and policy makers linked to La Vía Campesina and Coordinadora Latinoamericana de Organizaciones del Campo (Sevilla-Guzmán 2007). Agroecology has later been spread in Spain mostly as a social, anti-capitalist movement linked to food sovereignty principles, stronger settled down in the cities than in the countryside (López-García 2015). The interest from the academia and the public institutions towards agroecology has been really scarce, since agronomy in Spain is deeply rooted on a productivist ideology. In the last years, a bigger demand from local authorities has led to the development of different projects oriented to generate practical tools for what has been named as "political agroecology" (González de Molina 2013).

In 2012 a group of grassroots food movement activists in Spain organized a Seminar on "Local Agroecological Dynamization" with the aim of developing tools for promoting agroecological transitions within deagrarianized, postindustrial European territories. One of the results of the Seminar was to create a training program on the topic. The Universitat Autònoma de Barcelona grew interested in supporting a university post-graduate training program on LAeD. A multidisciplinary coordination team, including social and environmental researchers on agri–food systems and agri–food activists, developed an action-oriented program with the aim of training agroecological transition promoters. From a multi-actor perspective, targeted groups of students were both agri–food activists (including farmers) and public officers from local and rural development administrations. The first edition of the training program on LAeD was launched in 2014.

The three pillars of the course are: (1) an online training, (2) three meetings, and (3) the development of a research project. The 6 month online training consists of forum discussions based on the following theoretical and methodological contents: (1) rural sociology, agroecology and food sovereignty, (2) action-oriented social research and participatory methodologies, (3) alternative food networks and access to land, seeds and other resources, and (4) agroecological traditional knowledge. Additionally, practical exercises are included to confront theory with the perceived reality of students, along with the training program. Face-to-face meetings consist of three visits to different agroecological projects in different rural contexts. During these meetings (two full days each) participatory and qualitative research techniques are applied by the students in the context of the visited projects. The meetings include training sessions on facilitation skills through workshop simulations. Throughout the course, the students work on their final research project, which consists of developing a preliminary assessment of a real process, and designing a participatory process to address its prioritized challenges. Progress on the individual projects is collectively discussed on thematic groups during face-to-face meetings.

The training program is conceived as a hybrid forum generating action-reflection-action processes in two dimensions: (1) internal, in which the multi-actor network of students discusses niche–regime interactions, and (2) external, in which students have to generate such processes within their final project in real-world settings.

Data collection and analysis

The methodological design combined quantitative and qualitative methods: a survey questionnaire and in-depth interviews.

For a population universe based on 60 former students who took the postgraduate training program during 2 years (2014–2015 and 2015–2016) (2 courses × 30 students each course), a sample of 43 individuals answered an online questionnaire survey. The questionnaire was designed with the aim of studying the impact of the program, and it collected data on the individuals' perceptions about their daily work on local development projects. In addition to the classic socio-demographic questions (gender, age, educational background, and occupation), the questionnaire contained three sections: (1) individual motivations to enroll in the program, (2) perceived impacts (real or potential) on students' everyday work, and (3) employability of the individuals. The analysis consisted of univariate frequency analysis. In-depth interviews were performed with a sample of these same ex-students ($N = 16$) selected according to two main criteria: (1) type of institution or organization developing the projects (public or private), and (2) territorial level (i.e., municipal, county/province). Interviews were recorded, and interviewees were coded to guarantee confidentiality. The codes include reference to whether they work in the public sector (Pub), or in the private or non-profit sectors (Prv), and their institution/organization territorial level (county/province level: C, or municipal level: M). All of the interviewees were working in the agri–food and/or rural development fields when the survey was answered, and they were developing projects related to such issues (see Table 1). We followed Helen (2011) and coded relevant text passages from the voice records and notes using no predefined codes.

Potential impacts of LAeD training on individuals and institutions

The results were structured according to two types of the potential effects of the analyzed training program. The first related to the impacts on the individual level (i.e., local development practitioners themselves), and the second on the institutional level (i.e., the influence of the program in the local development institutions).

Reconfiguration of individual practices, concepts, and tools

The 43 individuals of our sample are involved in local and/or agri–food development as professionals or activists. The main motivations they mention for enrolling in the training program include: (1) to increase their theoretical knowledge on agroecology and/or food sovereignty (95%), (2) to increase their knowledge on participatory methods (81%) and other social research methods (68%), (3) and to connect with people and professionals related to these areas (71%).

Table 1 Topics, territories, and institutions related to projects developed by interviewees

Codes	Institutions (regions)	Projects
PubC1	Area Metropolitana de Barcelona (Catalunya)	Land planning regarding agricultural potential within Metropolitan Area of Barcelona
PubC2	Consell Comarcal del Vallès Oriental (Catalunya)	Local landraces, cooperative seed bank promotion
PubC3	Consorci del Lluçanés (Catalunya)	Prospection and promotion of local landraces crop and interchange networks
		'Local milk route' touristic development project, integrating farmers, tourism and catering sectors
PubC4	Local Action Group 'La Manchuela' (Castilla-La Mancha)	Set up and accompaniment of local network of farmers and consumers for development of local food networks
PubM1	St. Cugat del Vallés City Council (Catalunya)	Community Urban Gardens program participatory design
PubM2	València City Council (Pais Valencià)	Set up of Municipal Food Council for València City
		LAeD for agroecological transition in three Valencia City's agricultural districts
PubM3	Ea City Council (Euskadi)	Local development participatory strategy for BIZIASARE Association-Busturialdeko Garapen Iraunkorrerako Elkartea
PubM4	Orduña City Council (Euskadi)	Comprehensive dynamization of alternative food networks in Orduña County
PrvC1	Shepherd's School (Catalunya)	Specialized training itinerary design for new entrants into livestock farming
		Participatory research for local public slaughterhouse reopens in Pallars-Sobirá County
		'Espai test' program design—training on organic livestock farming and new entrants accompaniment
PrvC2	'Arran de Terra' Association-Collserola Natural Park (Catalunya)	Participatory process for agroecological transition in Collserola Natural Park
PrvC3	Association of Nine Town Councils and Five Civic Organizations (Euskadi)	Set up of County Food Council in Aiaraldea County
PrvC3	Producers' Group (Catalunya)	Participatory design of short agri–food chains
PrvC4	Livestock Farmers Association L'Esquella (Pais Valenciá)	Agroecological farming methods' advice for livestock farmers
		Participatory design of mobile slaughterhouse for organic pastoralist farmers
PrvM1	Private Consultancy for Cardedeu Town Council (Catalunya)	Dynamization of organic food networks (i.e., farmers' market) in Cardedeu Municipality
PrvM2	Private Consultancy (Catalunya)	Participatory process for agri–food sector in Canet de Mar Municipality, from an agroecological approach
PrvM3	Private Consultancy for Vallehermoso City Council and Garajonay National Park Consortium (Canary Islands)	Participatory design of training program for agroecological self-employment in Vallehermoso Municipality
		Participatory planning for fire prevention in Garajonay National Park
PrvM4	'La Troje' Association-Madrid Seeds Network (Madrid)	Participatory dynamization of Landrace Bean Growers Association of Madrid's Sierra Norte

Acquiring theoretical frameworks and methodological capacities and skills

Results show that students positively value the acquisition of new knowledge. Thus, students with a technical-scientific background tend to value knowledge related to social, political, and cultural aspects, and vice versa:

[I value] The social part, because of my background, is what I was lacking. The postgraduate course opened perspectives unknown for me, where I didn't have this base. And you have a point of departure. This has allowed providing me a theoretical framework to the process in [my village] (PubM4, background on agricultural engineering)

These are techniques applicable; I turn to the links especially on participatory methods, now I have a much broader range of participation tools (PubC4)

A majority of students claim to have covered new subjects after studying the training (51%) or expect to cover them shortly (61%), while 79% state that approaches and practices they were already familiar with have been reinforced. These results mean that most students were already familiar with the agroecological approach, to which the training has added new topics, and, most of all, it has provided them with greater confidence in its application, particularly within public employees (63%). At the same time, results show changes in students' attitudes and perspectives: 92% state to have modified their perspective on local and/or rural development, 85% state to have modified their perspective on what the role of public policy in the agri–food sector should include, and 74% state to have modified their perspective on the agri–food system. In all cases, changes have been made towards an agroecological and food sovereignty approach.

Students' perceptions of the training program differ according to their profiles. Public officers (more than the half of students) can be classified in two main groups: (1) those who had previous knowledge on the agroecological approach and asked permission from their institution to join the program, and (2) those who were encouraged (or obligated) by their institutions to join. While the first group expresses to be satisfied with the acquisition of methodological tools and a more systematic conceptual framework, the second group highlights the discovery of new approaches to agri–food systems, rural development, and the role of public administration on agri–food system sustainability. A third group, composed of food activists and organic farmers, acknowledged the reconfiguration of their perceptions about other agri–food system actors' roles and visions towards agroecological transition. This reconfiguration leads to the acquisition of a more complex perspective on agri–food system transitions, and therefore, raises the expectation for new methodological tools to be used in the actors' daily work.

Re-framing local development projects

Interviewees point out that several changes were triggered by their involvement in the projects, due in particular to the training, especially those regarding participatory approaches. They have applied some of the methods acquired in training (i.e., PubC3 and PubC2), as well as certain guidelines for the design of participatory processes. Such methods have been mentioned to be useful to generate mutual trust between actors in the territory, as well as to engage new actors in the process, generally adapting them to context specificities (i.e., PubC2, PrvM3, PrvC1, PrvC2). The practical cases and experiences visited during the training program have generated a greater self-confidence among students' practices, have helped them to identify key aspects of the processes in which they are involved, and have served as a comprehensive transition perspective for participatory processes design (i.e., PubM4, PubC4).

In some cases, acquired participatory methods have made it possible to redirect deadlocks in the transition processes, to increase agility and efficiency, and to better manage human relationships in the local development projects they are developing (PubM4). The idea of including a greater diversity of actors (e.g., the education community, food coops, extra-local or conventional farmers) in the processes is mostly repeated (PubC2, PubC3). The approach of the training program has opened communication channels amongst actors who previously did not participate in the processes or did not even mutually acknowledge themselves as valid interlocutors. In particular, this was the case for some local authorities and civic organizations (i.e., PrvC3):

I have seen that there are professionals in the public sector who are not familiar with the concepts of the training [program], but [I have seen] that they evolve... and this is why I have changed my perspective on the regional food council project, because I see there is technical staff in the region who can evolve... (PrvC3)

Most interviewees state their intention to engage new actors in future activities within their projects, as well as to reinforce the use of participatory methods (i.e., PubC2, PrvC2, PubC3, PubC4). Specifically, they mention the possibility of exploring the 'Campesino a Campesino' approach (PrvC1), a participatory diagnosis of the livestock sector (PrvC4), formal participatory bodies (e.g., steering group, monitoring committee) in charge of the territorial diagnosis and the elaboration of the action plan (PrvC2), or a participatory council with a diversity of actors (PubM2).

Influence on institutions

Emergent attitudes among institutional representatives

Interviewees perceive some changes in the positions of the public institutions involved in the training program or in those they work in. Indeed, increasingly positive attitudes towards agroecological projects are mentioned, which has the effect of generating more activities within this framework in the related territories (PubC4). Several cases have been mentioned in which participatory methods have generated great interest amongst the institutions promoting projects (i.e., PrvC2, PubM4). In some cases, the LAeD approach has provided a more structured methodological approach, which has been successfully applied in previously failed processes of local dynamization, and have been redirected in line with PAR and the agroecological transitions approach after the technical staff went through the training (i.e., PrvM1).

Interviewees also point out an increase in agroecological sensitivity amongst some political and technical staff members of both urban and rural local authorities (PubM1, PrvM2, PubM2), or they mention that at least interest for other points of view has been triggered amongst the institution promoting the projects (PubC1). It has also been stated that initiatives related to the training program have raised the interest of regional agrarian administrations (the Catalan Agriculture Department) for other approaches (i.e., agroecological) (PrvC1). An interesting finding is that public institutions interested in applying the agroecological and participatory approaches to their local development programs are fundamentally territorial (e.g., city councils, regional councils, provincial councils) but not directly agrarian. This is mainly due to the potential of the approach regarding local (territorial) administration competences over environmental conservation, territorial planning, city planning, economic promotion or employability. Sectorial agrarian institutions have shown themselves to be less open to adopting agroecological approaches.

New alliances or multi-actor network reconfigurations

Relationships between farmers and public administration officers seem to be improved through the training program approach (i.e., generating greater trust), especially in the case of initiatives not promoted by the administration or in the case of private technical assessment for the public sector (PrvC2, PrvM3, PrvM1, PrvC1). In one of these cases, it is explicitly stated that the LAeD approach has diminished the strong distrust previously existing between producers and the public administration (mainly due to previous experiences of top-down approaches to local development), through the set-up of new contexts and ways of communication and cooperation (PrvM3). Also, in several cases, new actors have engaged in the processes, either occasionally and/or informally, or permanently and/or formally. Only 39% of the questionnaire respondents state that, after their involvement in the training program, new alliances between social groups and/or the administration have occurred. Furthermore, 64% state that such alliances could be developed in the near futures, and 50% consider that already existing alliances have been reinforced.

The creation of alliances with new actors to develop joint actions is one of the most valued changes, both by the local authorities and by civic organizations (PubM2, PrvC3). Such new alliances are established with local administration representatives (PrvC3), schools (primary education) and other interested people in the region (PubC3). Additionally, the increase of the relationships between local authorities and other public administrations in the same region has been highlighted (PubM1, PrvC2, PrvM1), as well as that between private and social institutions such as food coops, between food coops and organic producers' groups (PubC4), and between the local authorities and farmers' groups. This last alliance has been mentioned in several cases as a remarkable change (PrvM3, PrvC1, PubM2, PrvC4). In one particular case (PubC2), the constitution of a participatory, food governance formal structure was referred to, in which several municipalities of the region enrolled together with producers and local associations to develop common goals and projects.

Local governments internal re-configurations

A case was reported in which the public administration explicitly included agroecology as a framework for local development in a regional center for economic and entrepreneurial promotion (PrvM1). Examples also exist in which the public administration sponsored a project for the creation of an agroecological agrarian extension mobile office (PrvM3). In other cases, a sector of the institution (e.g., department, council) undertook an agroecological approach and is trying to extend it to other areas in the same institution (e.g., town hall) (PubM2). Some institutions (i.e., Cardedeu and Sant Cugat del Vallés City Councils, Barcelona) have hired new staff as "Local Agroecological Dynamizators", to develop agroecological initiatives, justifying the positions for their potential to generate employment at the local level in a time of increased unemployment and social exclusion (due to the impact of the economic crisis) (PrvC2).

In several cases, interviewees claim that it is difficult for institutions to change because their structures are too rigid and unfavorable to new perspectives (PubM4, PubM1), or because the technical staff working with agri–food issues usually have such a background that their approach is very different than the agroecological one (PrvC1). In this sense, the interactions among students, and specifically public officers during peer-to-peer exchange served as a very useful tool to generate changes in the officers' perceptions of LAeD potential. In other cases, local authorities' officers appreciated the feasibility of applying the LAeD approach to their work. Conversely, local organic farmers linked to food movements did not see this possibility due to the strong resistance of the majority of local, conventional farmers to organic farming practices (PrvM3). Finally, it is also stated that the introduction of participatory procedures sometimes generates concern in the public institutions because it is seen as time consuming activity, and it involves changing current modes of operation (PubM1), thus generating internal resistance to change.

Discussion: agroecology training as 'hybrid forums' for regime reconfigurations?

We argue that the above described LAeD training program can be conceived as a hybrid forum, a convergence space where different actors meet, discuss and identify disruptive innovations within local agri–food systems. A 'hybrid forum' is the place where 'niche-actors' can be transformed into 'hybrid actors,' key actors in transferring niche innovation to the socio-technical regime level (which eventually would be reconfigured). Indeed, the feedback obtained from former students shows that the postgraduate training program had an impact not only regarding individual concepts, practices, and tools but also in their working environments (i.e., the institutions in which students work at the moment of the training or after).

According to our results, the LAeD teaching methodology is built upon action-reflection-action processes in which reversive approaches to the analyzed challenges are applied (Villasante 2006). Such analysis is made collectively from the students' own previous experience in conjunction with the new knowledge and experience generated during the training. This diverse learning space in which niche and regime-actors come together generates a learning process in which both typologies can be reconfigured. Thus new ways of mutual knowledge and recognition (of each other's challenges, values and positions) emerge which can be applied later in real life multi-actor networks dynamization.

Within the hybrid forum, students can set up creative interactions between niche and regime-actors to become hybrid actors and may act as innovation brokers (Galli and Brunori 2011). Some of the students of the training program become hybrid actors through a twofold process. First, through their final postgraduate thesis, students have to confront LAeD theories and methods with the real world in territories in which they usually work and/or live. Second, within the multi-actor network, they participate in action–reflection–action processes in which they confront their practical experiences on the field with other actors' values, visions and positions. Such twofold, collective action–reflection–action process acts, as stated by Freire (2004, 2005), as a way to learn to "read the World" with their own, innovative words at the same time that they learn "to read the word" (i.e., on participatory methods and sustainable agri–food systems). It is within this reconfiguration of the way students understand their work and their role within territories that they can learn how to tackle transition processes, in order to generate understanding among local actors, enhance capacities and support processes which lead to real learning and change (Woodhill 2009; Vogelezang et al. 2009; Darnhofer 2015). Thus, such shift leads students to occupy hybrid positions between niche and regime-actors to generate new niche–regime interactions which can lead to innovation anchoring and linking. Our results show several clear cases of rural development officers changing their views and starting new projects with agroecological orientation, as well as cases of agroecological activists changing their attitudes and accepting the possibility of cooperation with public services.

Also, hybrid forums have been identified as a key tool to develop new approaches to overcome previous frameworks which reproduce agri–food system agency inequalities (Smith 2007). Action–reflection–action processes within the program have led to the recognition of the role of small stakeholders (primarily farmers) as a kind of 'marginal actors' within local agri–food systems. Therefore these processes have uncovered the need for actions in which such actors get visibility and a stronger role, as recognized in previous research (Meek 2015; Méndez et al. 2017). Both participatory methods and contents such as traditional ecological knowledge or alternative food networks have been recognized with a strong potential to build up 'small actors' empowerment within agroecological transitions. In short, according to our results, LAeD training has been understood by students as an adequate tool to strengthen previous dynamics of agroecological transitions, through the improvement of their capabilities, self-confidence and applied knowledge.

As the training program is explicitly (but not only) oriented to local administration officers and it is officially associated with several local authorities (and it is offered by a public university), it may act as an institutional anchoring space (Elzen et al. 2012). Access to such institutional anchoring has been recognized as a high stage of socio-technical regimes reconfiguration (Bui et al. 2016). In this context, the training would represent an 'institutional innovation niche' by generating potential processes of niche anchoring and niche–regime linking. In this regard, the majority of the feedback collected from local administration officers shows a change in their viewpoints concerning the role of sustainable agrarian production for local development, in line with local agroecological proposals embedded in the territory. Some of the collected views are oriented towards the need to move from the rural development socio-technical regime focused in activities' diversification, which entails a deagrarianization dynamic (Marsden and Sonnino 2008; Guzmán et al. 2016), towards a new regime in which sustainable agrarian activity would be at the core of rural economic activity. Thus, we can say that through a training program based on participatory and multi-actor learning processes, the regime-actor nature of local administration officers can shift to positions of, at least, hybrid actors in agroecological transition processes.

Concerning agri–food systems, Wheeler (2008) identified agricultural administration officers as being a strong reluctant actor against sustainability transitions. Our results support this hypothesis showing that transition processes seem to be more difficult regarding agricultural administration officers. From the perspective of our theoretical framework, this may be because the agrarian administration is a key node for the reproduction of hegemonic regimes, and therefore, it could be situated as a regime-actor.

As for local administration officers, on the contrary, our research suggests they may act as formal hybrid actors. This role as hybrid actors is due to their twofold nature as niche and regime-actors (Elzen et al. 2012). It is also as a result of motivations and skills adopted to create new multi-actor alliances (niche network-anchoring) for agri–food systems sustainability (Roep and Wiskerke 2004; Elzen et al. 2012; Diaz et al. 2013). Public actors without direct competences on agriculture, such as local administrations and municipalities, would then be more exposed to become hybrid actors, and thus to integrate hybrid forums and catalyze niche–regime interactions through institutional anchoring. In this sense, the enrollment of local authorities (e.g., Àrea Metropolitana de Barcelona, Diputació de Barcelona, or several city councils) as sponsors within the training program represents a major opportunity. In fact, such institutions demand not only the training for their officers, but also the cooperation of LAeD students to develop innovative approaches for very specific challenges they face, such as the development of advisory and participatory processes on strategic land planning, or the design and implementation of agroecological transition processes.

The period analyzed is very short (only 2 years), and the sample is very limited, as our research has only exploratory purposes. In any case, it is possible to identify wider processes that provide a window of opportunity to push for socio-technical food regime reconfigurations, at least at the local scale (municipal or regional). Despite the chronic crisis of small actors involved in globalized agri–food systems, the recent crisis of local development itself (especially after the last decade of financial crisis and related austerity measures) is pushing local authorities (both urban and rural) to adopt and promote new approaches to local economies. As mentioned above, it is mainly local governments (city councils, county councils, provincial councils, metropolitan entities, etc.) that most tend to bet on alternative models of socio-economic development. These institutions are the first line in dealing with the social urgency caused by the economic crisis which led to the exclusion and impoverishment of many due to mass unemployment, evictions, energy shortages, food poverty, etc.

Although local governments tend to have fewer resources available, they are closer to the population and are taking on a leading role in managing the effects of the crisis. All of this creates a context in which agroecological initiatives can play an important role (Pomar-León and Tendero-Acín 2015), and some public administrations are beginning to consider them as useful formulas for improving social inclusion (e.g., job creation or lifestyle changes toward sustainability). As an example, there is an emergence of social gardens and other forms of urban and peri-urban agriculture (Domene 2016), along with increased social agriculture initiatives throughout the territory (Guirado et al. 2017). This emergence has been demonstrated by the projects promoted by former students of the training program.

The commitment to new local agroecological policies is much more intense in those municipalities where government teams have become aware of the importance of the environmental, energy and financial crisis (López-García et al. 2017) and are willing to look for other models of economic and social development. In this sense, local authorities need for innovative local development approaches are opening the socio-technical landscape to amplify the impact of the agroecological innovations generated by niche-actors. New alliances of agroecological research groups, food movements and non-agricultural local authorities are not at the center of food regime, but, perhaps, for this reason, they are more flexible and sensitive to regime reconfigurations and can become hybrid, bridge actors for agroecological transitions upscaling and outscaling.

Conclusions: deepening multi-actor networks role within agroecological transitions

In this paper we have suggested some insights on the way agroecological university training programs can influence food regime reconfiguration, focusing on multi-actor relationships from a MLP. With this aim, participatory methodologies have shown a high performative potential within the learning process itself.

The training program has also shown a high performativity on the reconfiguration of visions of local development and the different agri–food system's actors (e.g., the introduction of innovative narratives and action proposals within rural and local development practitioners and authorities). The set-up of training programs conceived as hybrid forums collecting an important diversity of agri–food system actors can be an effective tool to promote hybrid-actor profiles, both in students with an activist or farmer background, as well as in practitioners of local or agricultural administration or extension bodies. Within such hybrid forums, peer-to-peer interactions among the different actors act as an important change vector in a twofold way: (1) as a way to generate confidence among peers, and (2) as a way to confront different visions within the local agri–food systems.

The creation of hybrid forums is positive regarding such transitions, both at individual and collective levels. Action–reflection–action processes generate a resignification and reconfiguration of practices among students, which can be translated into their real-world environments, thus leading to innovation anchoring. The conformed hybrid actors act as a bridge between different positions and practices in the real world, facilitating the understanding and latter cooperation between different actors. This lead to collective innovation processes.

The methodological design of the LAeD training program addresses niche–regime interactions in different ways. On one side, providing both theoretically and methodologically specific tools oriented to the reconfiguration of visions on agri–food systems and multi-actor networks and focusing on social creativity processes can serve to overcome the lock-ins posed by regime-actors and regime-values against regime reconfiguration. On the other side, through a learning process within a mockup of the actors' diversity found in real transition processes to sustainable local agri–food systems, thus useful to experience the complex relations to be found in real world.

Therefore, the training offers a potential way to spread and accelerate niche-anchoring, at least, through a niche–regime linking along two different routes. Firstly, as a network anchoring tool generating directly and indirectly multi-actor networks, both along the course and through the final thesis developed on the real world. Secondly, as an institutional anchoring process as far as it includes agroecology in the toolkit of rural development officers and other practitioners. At the end, the training program may be understood as an "institutional innovation niche", as it explicitly promotes innovation processes within different public institutional bodies (both agricultural and territorial) which collaborate with the training program themselves. Thus, the training places agroecological innovations (both farming practices related issues and agroecological transitions methodological issues) at the core of the tension between regime-actors and niche-actors.

The transformative potential of the training program, in terms of niche–regime anchoring, is favored by the current crisis-related context of growing receptivity among different public bodies towards sustainability innovations within the agri–food system. The rise of some 'new municipalist' local governments in several Spanish big cities—such as Barcelona, València, Madrid, Zaragoza, Palma de Mallorca and others—and some regional governments—Comunitat Valenciana and Navarra-, with a novel and strong commitment to sustainable food policies opens a wide window of opportunity for sustainability transitions in the socio-technical landscape (López-García et al. 2017). Thus, addressing power issues through political action can become a strong support for such transitions, seeking for new alliances with those public bodies which could become sensible to agroecological transitions. The discussion on MLP applied to agroecological transitions can be therefore connected with the discussion on 'political agroecology', expressed as the need of "political and institutional mediation [...] (and) the development of public policies that drive the agroecological transition forward" (González de Molina 2013, p. 56). The way we have tried to link both issues—niche–regime interactions and political agroecology—is through (1) encouraging students to work on real-world processes, and (2) reinforcing dissemination activities of LAeD projects' results, with special regard to policy-makers.

The main feature that could differentiate this training from others is its marked hybrid forum character. It has been designed and promoted by a mix of academics and activists, and a great effort has been made to mix students working in local development institutions with students in the agroecological sector (activists and farmers). The teaching methodology encourages the interaction between student profiles with very different conceptual backgrounds. The training program is designed as a meeting place in such a way that it promotes intense interactions among the participants, generating bonds of trust between people who occupy key positions in their respective fields (public administrations, NGO activists of agroecology, farmers' cooperatives, etc.), but also provides them with real-world agroecological transition experiences, that turn some students into hybrid actors. The projects developed within the training, with the support of local institutions, have practical implications by implementing agroecological initiatives in the territories. However, we should be aware that hybrid forums can evolve both toward conventionalization of agroecology or toward the strengthening of its transformative potential. The impacts of the training open new possibilities to upscaling and outscaling agroecology, but this will not be promoted by hybrid actors alone, but mostly, as suggested by other scholars (González de Molina 2013; Levidow et al. 2014; Meek et al. 2017) by niche-actors and food movements, and reconsidering the current top-down agri–food policies.

Acknowledgements We would like to thank Ariadna Pomar-León and Guillem Tendero-Acín, as co-coordinators of the Training Program of Local Agroecological Dynamization, for their unvaluable cooperation for the present paper and in the training program itself. We would also like to thank all interviewees and respondants of the survey for their cooperation with the present research and their daily effort for agroecological transitions in Catalonia and Spain. Lastly, we would like to thank the special issue editors and the anonymous reviewers for their patient and invaluable comments, which really strengthened and enriched the paper.

References

Altieri, M.A. 1983. *Agroecología. Bases científicas para una agricultura sustentable*. Montevideo: Nordan Comunidad.

Bell, M.M., S.E. Lloyd, and C. Vatovec. 2010. Activating the countryside: Rural power, the power of the rural and the making of rural politics. *Sociologia Ruralis* 50 (3): 205–224.

Bell, S., and McAllister. 2012. *Sustainable agriculture through sustainable learning: Applying principles of adult learning to improve educational incomes*. Storrs, CT: Northeast SARE.

Bui, S., A. Cardona, C. Lamine, and M. Cerf. 2016. Sustainability transitions: Insights on processes of niche–regime interaction and regime reconfiguration in agri–food systems. *Journal of Rural Studies* 48: 92–103.

Bunch, R. 1985. *Dos mazorcas de maiz: una guía para el mejoramiento agrícola orientado hacia la gente*. Oklahoma City: World Neighbours.

Chambers, R. 1994a. Participatory rural appraisal (PRA): Challenges, potentials and paradigm. *World Development* 22 (10): 1437–1454.

Chambers, R. 1994b. The origins and practice of participatory rural appraisal. *World Development* 22 (7): 953–969.

Chambers, R. 1994c. Participatory rural appraisal (PRA)—Challenges, potentials and paradigm. *World Development* 22 (9): 1437–1454.

Cuéllar-Padilla, M., and Á. Calle-Collado. 2011. Can we find solutions with people? Participatory action research with small organic producers in Andalusia. *Journal of Rural Studies* 27 (4): 372–383.

Darnhofer, I. 2015. Socio-technical transitions in farming. Key concepts. In *Transition pathways towards sustainability in agriculture. Case studies from Europe*, eds. L.A. Sutherland, I. Darnhofer, G. Wilson, and L. Zagata, 17–32. Oxfordshire: CABI.

Darnhofer, I., W. Schneeberger, and B. Freyer. 2005. Converting or not converting to organic farming in Austria: Farmer types and their rationale. *Agriculture and Human Values* 22 (1): 39–52.

Diaz, M., I. Darnhofer, C. Darrot, and J.E. Beuret. 2013. Green tides in Brittany: What can we learn about niche–regime interactions? *Environmental Innovation and Societal Transitions* 8: 62–75.

Duru, M., O. Therond, and M. Fares. 2015. Designing agroecological transitions; a review. *Agronomy for Sustainable Development* 35 (4): 1237–1257.

Domene, E. (coord.). 2016. *Agricultura Metropolitana. Agricultura urbana i periurbana a l'àmbit metropolità de Barcelona: beneficis econòmics, socials i ambientals*. Barcelona: Institut d'Estudis Regionals i Metropolitans de Barcelona.

Elzen, B., F. W. Geels, C. Leeuwis, and B. van Mierlo. 2011. Normative contestation in transitions 'in the making': Animal welfare concerns and system innovation in pig husbandry. *Research Policy* 40: 263–275.

Elzen, B., B. van Mierlo, and C. Leeuwis. 2012. Anchoring of innovations: Assessing Dutch efforts to harvest energy from glasshouses. *Environmental Innovation and Societal Transitions* 5: 1–18.

Farrington, J., and A.E. Martin. 1988. Farmer participatory research: A review of concepts and recent fieldwork. *Agricultural Administration and Extension* 29 (4): 247–264.

Francis, C., G. Lieblein, S.R. Gliessman, N. Creamer, R. Harwood, L. Salomonsson, J. Helenius, D.H. Rickerl, R. Salvador, M.H. Wiedenhoeft, C. Flora, S.R. Simmons, M.A. Altieri, and R. Poincelot. 2003. Agroecology. The ecology of agri–food systems. Journal of Sustainable Agriculture 22 (3): 99–118.

Francis, C., T.A. Breland, E. Østergaard, G. Lieblein, and S. Morse. 2012. Phenomenon-based learning in agroecology: A prerequisite for transdisciplinarity and responsible action. *Agroecology and Sustainable Agri–food Systems* 37 (1): 60–75.

Franz, N. 2007. Adult education theories: Informing cooperative extension's transformation. *Journal of Extension* 45 (1). Available at http://www.joe.org/joe/2007February/a1.shtml.

Freire, Paulo. 1969. *¿Extensión o comunicación? La concientización en el medio rural*. Mexico, DF: Siglo XXI.

Freire, Paulo. 1975. *Pedagogia del oprimido*. Madrid: Siglo XXI.

Freire, Paulo. 2004. *La importancia de leer y el proceso de liberacion*. Mexico, DF: Siglo XXI.

Freire, Paulo. 2005. *Pedagogía de la esperanza: Un reencuentro con la pedagogía del oprimido*. Mexico, DF: Siglo XXI.

Galli, F., and G. Brunori. 2011. Knowledge brokerage to promote sustainable food consumption and production: Linking scientists, policymakers and civil society organizations. Report of the FOODLINKS project. http://www.foodlinkscommunity.net. Accessed 25 July 2017.

Geels, F.W. 2002. Technological transitions as evolutionary reconfiguration processes: A multi-level perspective and a case-study. *Research Policy* 31: 1257–1274.

Geels, F.W. 2010. Ontologies, socio-technical transitions (to sustainability), and the multi-level perspective. *Research Policy* 39: 495–510.

Geels, F.W. 2011. The multi-level perspective on sustainability transitions: Responses to seven criticisms. *Environmental Innovation and Societal Transitions* 1: 24–40.

Gliessman, S.R. 2002. *Agroecología: procesos ecológicos en agricultura sostenible*. Turrialba: CATIE.

Gliessman, S.R., and M. Rosenmeyer, eds. 2010. *The conversion to sustainable agriculture. Principles, processes and practices*. Boca Ratón: CRC Press.

González de Molina, M. 2013. Agroecology and politics. How to get sustainability? About the necessity for a political agroecology. *Agroecology and Sustainable Agri–food Systems* 37 (1): 45–59.

Guirado, C., N. Valldeperas, and A. Tulla. 2017. *L'agricultura social a Catalunya. Desenvolupament local i ocupació per a col·lectius en risc d'exclusió social*. Tarragona: Cossetània.

Guzmán, G.I., M. González de Molina, and E. Sevilla-Guzmán. 2000. *Introducción a la agroecología como desarrollo rural sostenible*. Madrid: Mundi-Prensa.

Guzmán, G.I., D. López-García, L. Román, and A.M. Alonso. 2013. Participatory action research in agroecology. Building organic food networks in Spain. *Agroecology and Sustainable Agri–food Systems* 37 (1): 37: 127–146.

Guzmán, G.I., D. López-García, L. Román, and A.M. Alonso. 2016. Participatory action research for an agroecological transition in Spain. In *Agroecology: A transdisciplinary, participatory and action-oriented approach*, eds. V.E. Méndez, C.M. Bacon, R. Cohen, and S.R. Gliessman, 140–160. Boca Ratón: CRC Press.

Heleba, D., V. Grubinger, and H. Darby. 2016. On the ground. Putting agroecology to work through applied research and extension in Vermont. In *Agroecology: A transdisciplinary, participatory and action-oriented approach*, eds. V.E. Méndez, C.M. Bacon, R. Cohen, and S.R. Gliessman, 177–192. Boca Ratón: CRC Press.

Helen, N. 2011. *Conducting research in conservation. Social science methods and practice*. London: Routledge.

Holt-Gimenez, E. 2008. *Campesino a campesino: Voces de Latinoamérica. Movimiento Campesino para la Agricultura Sustentable*. Managua: SIMAS.

Lamine, C. 2011. Transition pathways towards a robust ecologization of agriculture and the need for system redesign. Cases from organic farming and IPM. *Journal of Rural Studies* 27 (2): 209–219.

Levidow, L., M. Pimbert, and Y.G. Vanloqueren. 2014. Agroecological research: Conforming or transforming the dominant agro–food regime? *Agroecology and Sustainable Agri–food Systems* 38: 1127–1155.

Lieblein, G., T.A. Breland, C. Francis, and E. Østergaard. 2012. Agroecology education: Action-oriented learning and research. *Journal of Agricultural Education and Extension* 18 (1): 27–40.

Lobley, M., A. Butler, and M. Reed. 2009. The contribution of organic farming to rural development: An exploration of the socio-economic linkages of organic and non-organic farms in England. *Land Use Policy* 26: 723–735.

López Vargas, G., M. Ardón, and E. Tomás Pérez. 2009. *Agroecología práctica*. Tegucigalpa: LITHOCOM.

Lopez-Garcia, D., and G.I. Guzman Casado. 2014. *Metodologias participativas para la transicion agroecologica*. Madrid: Sociedad Española de Agricultura Ecologica.

López-García, D. 2015. *Producir alimentos, reproducir comunidad. Redes alimentarias alternativas como formas económicas para la transición social y ecológica*. Madrid: Libros en Acción.

López-García, D., L. Calvet-Mir, J. Espluga, M. Di Masso, G. Tendero-Acin, and A. Pomar-León. 2015. La dinamización local agroecológica como estrategia para la construcción de soberanías locales. *Ecologia Politica* 49: 28–34.

López-García, D., P.M. Herrera, N. Alonso, J. Mérida, and J.M. Pérez. 2017. 'Cities for agroecology' networks in Europe and Spain. *Urban Agriculture* 33: 55–57.

Marsden, T., and R. Sonnino. 2008. Rural development and the regional state: Denying multifunctional agriculture in the UK. *Journal of Rural Studies* 24: 422–431.

Meek, D. 2015. The cultural politics of the agroecological transition. *Agriculture and Human Values* 33 (2): 275–290.

Meek, D., K. Bradley, B. Ferguson, et al. 2017. *Agric Hum Values*. https://doi.org/10.1007/s10460-017-9780-1.

Méndez, V.E., C.M. Bacon, and R. Cohen. 2016. Introduction: Agroecology as a transdisciplinary, participatory and action-oriented approach. In *Agroecology: A transdisciplinary, participatory and action-oriented approach*, eds. V.E. Méndez, C.M. Bacon, R. Cohen, and S.R. Gliessman, 1–22. Boca Ratón: CRC Press.

Méndez, V.E., M. Caswell, S.R. Gliessman, and R. Cohen. 2017. Integrating agroecology and participatory action research (PAR): Lessons from Central America. *Sustainability* 9 (5): 705. https://doi.org/10.3390/su9050705.

Milestad, R., R. Bartel-Kratochvil, H. Leitner, and P. Axmann. 2010. Being close: The quality of social relationship in a local organic cereal and bread network in Lower Austria. *Journal of Rural Studies* 26 (3): 228–240.

Neumeier, S. 2012. Why do social innovations in rural development matter and should they be considered more seriously in rural development research? Proposal for a stronger focus on social innovations in rural development research. *Sociologia Ruralis* 52 (1): 48–69.

Ota, C., C.F. DiCarlo, D.C. Burts, R. Laird, and C. Gioe. 2006. Training and the needs of adult learners. *Journal of Extension* 44 (6). Available at http://www.joe.org/joe/2006december/tt5.shtml.

Padel, S. 2001. Conversion to organic farming: A typical example of the diffusion of an innovation? *Sociologia Ruralis* 41 (1): 40–61.

Padel, S. 2008. Values of organic producers converting at different times: Results of a focus group study in five European countries. *International Journal of Agricultural Resources, Governance and Ecology* 7 (1–2): 63–77.

Ploeg, J.D., and T. Marsden. 2008. *Unfolding webs. The dynamics of regional rural development*. Assen, NL: Van Gorcum.

Pomar-León, A., and G. Tendero-Acín. 2015. *Ja Volem el Pa Sencer. Respostes a la pobresa alimentària en clau de Sobirania Alimentària*. Barcelona: ASAC!.

Reed, M. 2008. The rural arena: The diversity of protest in rural England. *Journal of Rural Studies* 24 (2): 209–218.

Rhoades, R.E., and R. Booth. 1982. Farmer-back-to-farmer: A model for generating acceptable agricultural technology. *Agriculture Administration* 11: 127–137.

Roep, D., and J.S.C. Wiskerke. 2004. Reflecting on novelty production and niche management in agriculture. In *Seeds of transition*, eds. J.S.C. Wiskerke and J.D. van der Ploeg, 341–356. Assen, NL: Van Gorcum.

Rogers, E.M. 1962. *Diffusion of innovations*. New York: The Free Press.

Röling, N.G., and M.A.E. Wagemakers, eds. 1998. *Facilitating sustainable agriculture: Participatory learning and adaptive management in times of environmental uncertainty*. Cambridge: Cambridge University Press.

Sanderson, B.A.A., and A.A.R. Ioris. 2017. Addressing the knowledge gaps in agroecology and identifying guiding principles for transforming conventional agri–food systems. *Sustainability* 9: 330. https://doi.org/10.3390/su9030330.

Sevilla-Guzmán, E. 2007. *De la sociología rural a la agroecología*. Barcelona: Icaria.

Smith, A. 2007. Translating sustainabilities between green niches and socio-technical regimes. *Technology Analysis and Strategic Management* 19 (4): 427–450.

Tisenkopfs, T., G. Brunori, K. Knickel, and S. Sumane. 2009. Co-production of rural innovation: Towards an enriched theoretical model. In *Innovation processes in agriculture and rural development. Results of a cross-national analysis of the situation in seven countries, research gaps and recommendations*, eds. K. Knickel, T. Tisenkopfs, S. Peter, and S. Sumane, 129–159. INSIGHT project. http://www.insightproject.net. Accessed 10 June 2017.

Ventura, F., G. Brunori, P. Milone, and G. Berti. 2008. The rural web: A synthesis. In *Unfolding webs, the dynamics of regional rural development*, eds. J.D. Ploeg, and T. Marsden, 149–174. Assen, NL: Van Gorcum.

Villasante, T.R. 2006. *Desbordes creativos*. Madrid: La Catarata.

Vogelezang, J., A. Wals, B. van Mierlo, and F. Wijanads. 2009. Learning in networks in Dutch agriculture: Stimulating sustainable development through innovation and change. In *Transitions towards sustainable agriculture and food chains in peri-urban areas*, eds. K. Poppe, C. Termeer, and M. Slingerland, 93–111. Wageningen, NL: Wageningen Academic Publishers.

Wezel, A., H. Brives, M. Casagrande, C. Clément, A. Dufour, and P. Vandenbroucke. 2015. Agroecology-territories: Places for sustainable agriculture and agri–food systems and biodiversity conservation. *Agroecology and Sustainable Agri–food Systems* 40 (2): 132–144.

Wheeler, S.A. 2008. What influences agricultural professionals' view towards organic agriculture? *Ecological Economics* 65 (1): 145–154.

Woodhill, J. 2009. Institutional innovation and stakeholder engagement. Linking transition management in the North with development in the global South. In *Transitions towards sustainable agriculture and food chains in peri-urban areas*, eds. K. Poppe, C. Termeer, and M. Slingerland, 273–291. Wageningen: Wageningen Academic Publishers.

Daniel López-García (Ph.D. in Agroecology) is a researcher at Fundación Entretantos, coordinator of the Post-graduate Program 'Local Agroecological Dynamization' of the Universitat Autònoma de Barcelona, and lecturer at three master programs on Agroecology and Organic Farming. His major research field is agroecological transitions, focusing on alternative food networks, localized food systems, agroecology-oriented food policies and food governance. He has combined mixed research approaches combining participatory, qualitative and quantitative social research. Recent research focuses on participatory methodologies for urban and city-region food policy planning.

Laura Calvet-Mir (Ph.D. in Environmental Science) is a research fellow at IN3-UOC (Barcelona, Spain) and affiliated researcher at ICTA-UAB (Barcelona, Spain). Her research focuses on urban and rural agriculture, agroecological transitions and biocultural diversity. In studying local social-ecological systems, Dr. Calvet-Mir has combined ethnographic approaches with quantitative methods such as statistics and social network analyses. Her main research interests are agrobiodiversity, agroecology, ecosystem services, ethnoecology, political ecology, social network analysis, and traditional ecological knowledge.

Marina Di Masso Tarditti (Ph.D. in Sociology) is a Juan de la Cierva research fellow at the Agroecology and Alternative Food Systems Chair of the UVic-UCC, Spain. Her major research interests are transformative processes of social change. Particularly her research focuses on critical perspectives of the global agrifood system, alternative food networks, and food sovereignty as a transformative paradigm. Recent research and personal interests include wider alternative socioeconomic practices, with a focus on feminist economics and social and solidarity-based economy. She is a member of the Inclusive Societies, Policies and Communities at Uvic-UCC.

Josep Espluga is a sociologist, associate professor at the Department of Sociology at the Universitat Autònoma de Barcelona (UAB) (Spain), and researcher at the Institute of Government and Public Policies (IGOP). He is also member of the Research group for Agriculture, Livestock and Food under Globalisation (ARAG-UAB). His major research field is environmental and health sociology, with special focus on the social perception of risk. He is co-director of the Post-graduate Program 'Local Agroecological Dynamization', and coordinates the Minor on Sustainable Development and Global Citizenship of the UAB.

Agriculture and Human Values (2019) 36:581–594
https://doi.org/10.1007/s10460-018-9870-8

SYMPOSIUM/SPECIAL ISSUE

What's wrong with permaculture design courses? Brazilian lessons for agroecological movement-building in Canada

Marie-Josée Massicotte[1] · Christopher Kelly-Bisson[2]

Accepted: 2 June 2018 / Published online: 26 June 2018

Abstract

This paper focuses on the centrality of permaculture design courses (PDCs) as the principal sociopolitical strategy of the permaculture community in Canada to transform local food production practices. Building on the work of Antonio Gramsci and political agroecology as a framework of analysis, we argue that permaculture instruction remains deeply embedded within market and colonial relations, which orients the pedagogy of permaculture trainings in such a way as to reproduce the basic elements of the colonial capitalist economy among its practitioners. In the specific case of eastern Ontario, this embeddedness had the effect of diluting the transformative capacity of permaculture practitioners who were unable to create its own social movement organization. The paper then highlights key elements of the agroecological pedagogy used by the Brazilian Landless Rural Workers Movement (MST) and the *Escola Latinoamericana de Agroecología* (Latin American School of Agroecology, or ELAA) in Paraná, Brazil. The objective is to draw lessons from these inspiring experiences, in a rather unique context of struggles that can help to critically assess the pedagogical practices and principles presently informing permaculture communities in Canada and in advanced industrialized countries more generally. We then conclude by reiterating the key arguments and lessons drawn from the Brazilian pedagogical experiences, pointing out the importance of engagement and coalition-building with established rural and urban movements, as well as progressive farmer, Indigenous, and rural associations to foster a just and sustainable transformation of agri-food systems, starting at the local and regional levels. It also emphasizes the need for the most marginalized sectors to lead the way towards an agroecological transition.

Keywords Permaculture · MST · Agroecology · Emancipatory pedagogy · Peasant movements · Gramsci

Abbreviations

CPT	*Comissão Pastoral da Terra* (Pastoral Land Commission)
CUFF	Community Urban Food Forest
ELAA	*Escola Latinoamericana de Agroecología* (Latin American School of Agroecology)
MST	*Movimento dos Trabalhadores Rurais Sem Terra* (Landless Rural Workers Movement)
NCC	National Capital Commission
OPIRG	Ontario Public Interest Research Group
PDC	Permaculture Design Course
PIEO	Permaculture Institute of Eastern Ontario
PO	Permaculture Ottawa
PRI	Permaculture Research Institute

We have recently witnessed the multiplication of analyses criticizing the Green Revolution originating from initial critics like Carey McWilliams (1939), Murray Bookchin (1962), and Rachel Carson (1962) as a result of the intensification and expansion of industrial agriculture practices. Scholars such as Phil McMichael (2009) and Haroon Akram-Lodhi (2013) have highlighted the multiple and pervasive impacts of such large-scale agri-food production on biodiversity, the environment, as well as the quality of life and well-being of food producers, consumers, future generations, and non-human species alike. It is in this context that a growing number of food producers, ecologists, indigenous, and women's movements have been organizing and promoting food sovereignty, agroecology, and food justice, especially since the inclusion of agriculture in the governance structure of the World Trade Organization in 1995. These sociopolitical

✉ Marie-Josée Massicotte
massicot@uOttawa.ca

1 School of Political Studies, Institute for Feminist and Gender Studies, Ottawa, ON, Canada

2 Political Science, School of Political Studies, University of Ottawa, Ottawa, ON, Canada

actors are seeking ways to challenge the globalizing tendency of this dominant model of mostly monocultures and to promote socially and environmentally more sustainable alternatives. Permaculture is one such example of a holistic, ecosystem-based design practice that focuses on transitioning away from conventional agricultural production to create sustainable community living spaces (Holmgren 2002; Mollison 1988).

In this article, we focus on the centrality of permaculture design courses (PDCs) as the principal strategy of the permaculture community in Canada to transform local food production practices. Building on the work of Antonio Gramsci and using political agroecology as a framework of analysis, we argue that permaculture instruction has been and remains deeply embedded within market and colonial relations, which orient the pedagogy of permaculture trainings in such a way as to reproduce the basic elements of the colonial capitalist economy among its practitioners.[1] In the specific case of eastern Ontario[2]—focused on in this paper—this embeddedness had the effect of diluting the transformative capacity of permaculture practitioners who were unable to create its own social movement organization. It eventually resulted in its dissolution and reorganization, since 2013, as a network of market producers and trainers.

The paper first reviews the historical origins of permaculture, as well as its political economic and pedagogical practices, which became globally manifested in the form of the permaculture design courses (PDCs). Secondly, the paper introduces the Gramscian theoretical framework from which we draw to analyze the strategic, political, and organizational challenges faced by the eastern Ontario permaculture community, especially as it relates to the delivery of PDCs. We discuss the role of PDC instructors as "organic intellectuals" who contribute to a "war of position". These key Gramscian concepts, as we will argue, are useful for understanding the pedagogical capacity and limitations of permaculture advocates in their mobilization of food producers through social and ecological consciousness in a struggle against the precarious and alienating nature of food production in Canada. The third section of the paper details a case study of the eastern Ontario permaculture community to identify the socioeconomic and cultural dilemmas and constraints experienced by permaculture trainers and participants, as they are embedded in a largely market-based country such as Canada or the USA. We explore how this particular positioning works against the broader sociopolitical changes that permaculture advocates hope to accomplish. The fourth section highlights key elements of the agroecological pedagogy of the *Movimento dos Trabalhadores Rurais Sem Terra* (Landless Rural Workers Movement, hereafter MST) and the *Escola Latinoamericana de Agroecología* (Latin American School of Agroecology, or ELAA) in Paraná, Brazil. The objective here is to draw lessons from these inspiring experiences, in a rather unique context of struggles that can help to critically assess the pedagogical practices and principles informing permaculture communities in Canada and in advanced industrialized countries more generally. We then conclude by reiterating the key arguments and lessons drawn from the Brazilian pedagogical experiences in financially poor agricultural communities. Our analysis of these various praxes thus points out the importance of engagement and coalition-building with established rural and urban movements, as well as progressive farmer, Indigenous and rural associations to foster a just and sustainable transformation of agri-food systems, starting at the local and regional levels. It also emphasizes the importance of organic intellectuals from most marginalized sectors to lead the way towards an agroecological transition, whereas PDCs trainers have tended to function as technical trainers from affluent social backgrounds without an explicit intention of participating in a transformation of agri-food systems.

This paper draws from extensive field work research conducted in eastern Ontario, Canada. The field work involved a full year of participatory action research by Christopher Kelly-Bisson (2013) for the completion of the Master of Arts thesis. From its foundation in 2011, he was an active organizer with Permaculture Ottawa (PO); Kelly-Bisson documented the regular activities and gained insightful access to participants, attaining a good understanding of the tensions at play within such organizations. The case study research also entailed ten semi-structured, in-depth interviews conducted over a full year of involvement within the eastern Ontario permaculture community in 2011 and 2012. The field work included the completion of a PDC with the Permaculture Institute of Eastern Ontario (PIEO). As an organizer, Kelly-Bisson also got involved in the experiment to establish a non-profit Community Urban Food Forest (CUFF) on one acre of land owned by the Government of Canada.

Marie-Josée Massicotte (2014) conducted field research on the MST and the Latin American School of Agroecology (ELAA) in Brazil, using qualitative methods to conduct a discourse analysis of agroecological trainings and food sovereignty as a mobilizing concept. She completed over 20

[1] Gender and racial issues are particularly relevant to analyze both in the Canadian and Brazilian contexts, but for the purpose of this paper we mostly focus on the effects of a permaculture pedagogy that is embedded in a neoliberal capitalist and colonial context, from a Gramscian perspective. Nonetheless, we recognize that an intersectional approach would be extremely useful to make visible gender and racial inequalities that persist among agroecology and permaculture communities.

[2] Eastern Ontario is not a formal political designation. For our purpose, and as it is used in this particular region, it covers the area in Ontario east of Toronto and south of Deep River.

semi-structured interviews, between one and two and a half hours long, during six field work periods that lasted 10–30 days, between 2005 and 2010. Interviewees were mostly leaders and grassroots activists of the MST and participants at the ELAA school. Numerous informal conversations with ELAA activists-learners-farmers, activists-researchers-teachers and a few allies were also essential to appreciate the pedagogical context and the daily activities at the ELAA school. The opportunities to attend the 2008 *Jornada de Agroecología*, the 2009 World Social Forum in Belém, Brazil, as well as various seminars and workshops during three different visits at the ELAA school further deepened the researcher's capacity to make sense of the pedagogy and functioning of the school and the interdependencies between rural movements, state authorities, as well as local, regional and transnational solidarity networks. These insights will be used to compare and draw lessons that can be useful to understand the limitations faced by the permaculture community in eastern Ontario, Canada.

The origins, political economy, and pedagogy of permaculture

> The Prime Directive of Permaculture: The only ethical decision is to take responsibility for our own existence and that of our children. Make it now.
> The Principle of Cooperation: Cooperation, not competition, is the very basis of existing life systems and of future survival (Mollison 1988, pp. 1–2).

Few academic studies have been conducted on permaculture design and its community of practitioners globally, potentially because of the harsh criticism of the holistic approach developed by Bill Mollison for not being easily amenable to "scientific" experimentation (Ferguson and Lovell 2014; Veteto and Lockyer 2008). Most of the dedicated research on permaculture has focused on specific practices with little consideration of its practitioners as political actors (Ferguson and Lovell 2017a, b; Morel et al. 2016; June Brawner 2015; Kennedy 1991; Jungck 1985). For example, Veteto and Joshua (2008) and Aiken (2017) have discussed the political implications of permaculture within other movements, but these analyses miss the opportunity to explore the politics at the core of permaculture thought and action. Documentation of the ecovillage movement in the United States (Dawson 2006; Bang 2005) and the transition towns movement in the United Kingdom (Hopkins 2008) address their underpinning political positions, however these movements emerged more out of Green Party or deep ecology politics and tend to be only peripherally informed by permaculture practices. Puig de la Bellacasa (2010) conducted a very useful conceptual analysis through her participation in a PDC on the feminist care ethics of permaculture. Nonetheless, this study did not document particular practices of political engagement. For their part, Ferguson and Lovell (2015) conducted a much-needed demographic profile of the participants in the global permaculture network, which they describe as a social movement. However, they use the term "permaculture movement" without detailing any of its specific structure or political strategies, leaving it unclear how such a movement is to be defined in relation to the literature on social movements.

This article will make the case that what are largely white, middle-class permaculture communities do not qualify as a social movement as defined through a Gramscian perspective involving a coherent consciousness with a discernible praxis for sociopolitical, economic and cultural changes. In this sense, it seeks to fill the gap identified by Ferguson and Lovell (2015) who call for further analysis of the political strategy and actions of the global permaculture network promoting a sustainable agri-food transition. Ultimately, the existing literature has been successful in detailing and analyzing the practices and demographics of permaculture communities. However, it has yet to give greater attention to their pedagogy and their specific political context, strategies, and outcomes. Hence, this contribution provides such a political analysis by examining the experience of the eastern Ontario permaculture community. Our findings underline the risk that permaculture communities composed of middle-class practitioners reinforce neoliberal market relations in their social relationships. This is in-part the result of the pedagogical structure and central role of the PDC that emphasizes the importance of professional expertise and its application through individual entrepreneurship mostly on private property which, as we will highlight later, contrast with the Brazilian experiences of the MST and the Latin American School of Agroecology.

Permaculture emerged as an ecosystems-based design technique in Australia in the 1970s and was developed by the interdisciplinary science instructor Bill Mollison and ecological design student David Holmgren. This design technique combined insights from the literature on "natural farming" developed by Masanobu Fukuoka (1990), agroforestry practices by Smith (1950), and ecosystem theory by Howard Odum (1971). The imperative for the development of permaculture was based on the desire to create self-sufficient and sustainable communities in a time where ecological crisis and threats of nuclear war were escalating (Conford 2011; Holmgren 2011, p. 23; Veteto and Joshua 2008). As such, permaculture included an acknowledgement of the need for individuals to take personal responsibility for environmental destruction and for adopting autonomous, self-sufficient practices. Mollison and Holmgren began by documenting their backyard experimentations in Tasmania, as part of Holmgren's undergraduate thesis, which then was fully developed into their seminal, co-authored book *Permaculture One*, in 1978. Mollison immediately set out to develop a curriculum of popular education to spread the design practice internationally (Smith 2011). He offered the first PDC in 1981 (Holmgren 2011, p. 23). Throughout the 1980s, Mollison taught PDCs to thousands of people, some of whom went on to write books that are now at the core of permaculture literature. In 1988, Mollison published the now classic *Permaculture: A Designers' Manual*, which compiled the main elements of the PDC curriculum and its pedagogic principles. By this time, permaculture trainings had become much more standardized so that students could independently recreate and promote this design system around the world. Hence, by the late 1980s, permaculture was a global network of instructors and consultants based on a specific method of ecological design building mostly on Mollison's work. Nonetheless, because of the principles behind permaculture itself, the training method remains flexible enough to adapt to various ecosystemic conditions (tropical, arid, temperate regions, see Smith 2011).

The global permaculture network is centered on the delivery of PDCs, consisting of 72–84 h of intensive trainings which cost on average between 1000 and 2000 USD (Mann 2012). A few institutions and instructors—such as the Permaculture Institute of Eastern Ontario—have offered peer-subsidized, sliding scale, or free tuition for the courses. In most cases, PDCs are delivered intensively, with students living on-site-at a permaculture operation-for the entire duration of the course, but there are also options for series of weekend retreats. PDCs consist of mostly hands-on ecological design training for individual learners, with practical elements of landscape architecture, natural resource management, and regional planning, but they also incorporate permaculture theory, didactic instruction, and individual study (Smith 2011; Mollison 1988). Permaculture adepts continue to grow and promote inclusive objectives, based on a model of "train-the-trainer" educational retreats, as well as social cooperation and sustainability. However, this global network has not yet reached a scale or capacity to make significant changes in land use and agricultural policies and practices (Ferguson ad Lovell 2014).

PDCs are not officially recognized in most places by non-permaculture institutions, and the leaders among the global permaculture community are still debating whether having a PDC even "certifies" one to be a permaculture designer (PIEO personal correspondence with author June 3, 2012). Australia's Permaculture Research Institute (PRI), which was founded by Bill Mollison and is currently led by Geoff Lawton, was established to ensure the quality and consistency of knowledge and instruction for permaculture teachers and certification. Mollison encouraged his students to deliver PDCs around the world as the program became more structured and standardized. The PRI therefore can verify who was given teaching certification through certified permaculture institutes (Lawton 2011). However, attempts by institutions such as the PRI to officially regulate the delivery of PDCs have been significantly limited. There is now a wide gap in terms of what people perceive as a legitimate certification in permaculture design, and if PDCs can or should result in a certification at all (PIEO personal correspondence with author, June 3, 2012). Regardless of the institutional ambiguity of the PDC, it still holds a great deal of informal authority and many within the global network are actively working on innovative methods that will most probably continue to shape pedagogical practices and PDCs.

The PDC, in its self-regulated capacity, functions among permaculture practitioners as a means of ensuring alignment of practices and trainings with Mollison's specific vision of permaculture. A first tension thus emerges between the desire to disseminate permaculture training broadly and the ability of the global network to maintain the quality of its curriculum. Until now, this tension is mainly resolved through the specialization of permaculture trainers—by having practitioners increasingly dependent on delivering training for their living—relying upon a self-help style of workshop. In this model, students and teachers engage in a relationship of commodity exchange that is independent of formal institutions of higher education, trading a relatively costly tuition fee for sharing and acquiring the skills deemed necessary to be recognized as a qualified permaculture specialist worldwide. As the following case study of the eastern Ontario permaculture community will demonstrate, the embeddedness of trainers within a market economy and a deeply consumerist and individualist society—which also marked many other regions—raise important challenges and dilemmas for permaculture advocates.

Seeing post-capitalist experimentations through Gramsci's war of position

Antonio Gramsci (1891–1937) was a founding member of the Italian Communist Party. His political theory draws upon his experience as a revolutionary organizer detailed in a series of notebooks written in prison between 1926 and 1937, under the fascist regime. Multiple researchers, such as Karriem (2013) and Motta (2017), have analyzed his influence in Latin America and especially among Brazilian landless rural workers and peasants, to think about social changes and to promote the deployment of critical pedagogies. Gramsci's work remains a useful source for exploring the transformative role of pedagogy among the working class to foster an emancipatory political project, or what he called a "counter-hegemonic movement"[3] capable of contesting the dominant "common sense". The latter concept refers to the prevailing values and ideology that contribute to make the imperial-capitalist world order acceptable to the majority, including less powerful subaltern forces. Gramscian analysis recognizes that every person is an "intellectual" who possesses the faculties to perceive their world self-consciously, to form opinions, desires, and critical reasoning about the best ways of organizing everyday life. Gramsci equally asserts that only some people assume intellectual roles: "all men [sic] are intellectuals... but not all men have in society the function of intellectuals" (Gramsci and Hoare 1971, p. 9). In this sense, Gramsci allows us to contest the Western tendency to exclude popular knowledges emerging from people's daily practices and interactions with the natural world, including from peasant experiences.

Gramsci further specifies the existence of "organic intellectuals" who are activists emerging from within a particular class—bourgeois or working class—and assuming the role of "specialized" intellectuals; they are recognized by their peers for their ability to foster a sense of self-awareness among members of their class (Gramsci and Hoare 1971, p. 5). Whereas traditional intellectuals are "co-opted" to defend and diffuse the interests and values of elite forces and liberal democratic institutions, thus contributing to the acceptance of the bourgeois hegemony as if it benefits all classes (pp. 6–7), organic intellectuals can emerge from lower-middle classes and marginalized groups. In such cases, these intellectuals actively participate in labor production and represent workers' interests (Gramsci and Hoare 1971, p. 10). They gain a critical awareness through their everyday experiences of oppressive structures and injustices, as well as through interactions with their physical environment and with members of their class. This "coming-to-awareness" is part of a collective political process to make visible power dynamics and to "develop the capacity to think and engage in autonomous political activity" (Karriem 2013, p. 143). By diffusing alternative values and interests of working class communities, organic intellectuals thus participate in what Gramsci identified as a "war of position" that can lead to the emergence of counter-hegemonic movements (Morton 2007). Applied to today's context, such movements can eventually displace the neoliberal and colonial state and the capitalist interests and white privileges that it supports.

The other Gramscian concepts that we are drawing from are the "war of position" within the "extended state", a war that is waged by activists towards the gradual transformation of power relations and societies. In our study, the war of position includes the politicization of permaculture within permaculture networks, and the acknowledgement of ecology and justice as central elements of agricultural production and livelihoods throughout broader social movements. Gramsci's understanding of the "extended state" is useful here as it includes civil society plus political society (Hoare and Gramsci 1971, pp. 12–13). Political society refers to formal political institutions, such as governments, the army, the legislative and administrative branches of the state, and the judicial system. Civil society includes schools, religious and civic organizations, as well as mass media, the sphere within which the war of position is waged to gain the consent of the majority. As Leandro Vergara-Camus maintains, the capitalist culture of dominant classes is diffused within civil society: "Ultimately, bourgeois hegemony rests on what Gramsci calls the 'common sense' of society, which consists of the values, norms and institutions that are so internalized that they are understood as natural and organize our daily life" (Vergara-Camus 2014, p. 93). Gramsci believed that an emancipatory project should emerge from marginalized communities becoming conscious of the multiple oppressions and organizing themselves.

Among contemporary urban and rural working classes, some organic intellectuals are mobilizing and politicizing within their respective communities to foster sociopolitical and cultural changes, from the bottom-up. Contrary to orthodox Marxist analyses, Gramsci's work offers a set of strategies for transforming dominant structures of power and "collective will", including popular education and politicization among marginalized communities. In contrast to traditional educators who train individuals to "consent" and respond to market's needs, we are witnessing the emergence of a growing number of agroecological movements and trainers who privilege a mutual learning model that focuses on political organizing and knowledge-building based on ecologically and culturally adapted practices. Such grassroots movements open spaces and invite citizens to participate in counter-hegemonic struggles. In a similar way, our analysis highlights how permaculture communities can learn from agroecology as a politicized pedagogical model that contributes to challenge today's neoliberal agri-food regime, as well as the liberal capitalist preferences for private property and consumerism.

[3] In Gramsci's work, hegemony involves both consent and coercion; to maintain their power, the rulers, or dominant forces, need to build alliances with other classes and convince a majority that the prevailing order is also beneficial for them, thus obtaining their consent. However, hegemony is never complete. The constant presence of resistance forces requires the availability of coercive means to dissuade lower classes, the "ruled", to rebel and form a counter-hegemonic movement.

The permaculture community of eastern Ontario, Canada

The permaculture community of eastern Ontario existed in an organized and self-conscious capacity between 2011 and 2013, a 2-year period during which Kelly-Bisson was fully involved in its activities. Several leaders among this community drew from their experiences with other radical groups in the city, and especially from Occupy Ottawa, which represented a prefigurative mode of living and organizing that explicitly contested capitalism, colonialism, sexism, and other forms of oppression (Gordon 2008). Participation of these leaders in Occupy Ottawa and similar movements informed the initial purpose of the community to apply permaculture design practices to establish just and equitable livelihoods, as an alternative to the neoliberal and colonial order. However, the permaculture community eventually dissolved into a network of individual entrepreneurs organized through market exchange contrary to its initial post-capitalist vision. It resulted in the exclusion of many of its members without the capacity to start their own businesses. This section details the structural constraints and context that led to this outcome. It also highlights the specific role of PDCs and its commodity production of expertise, which reinforced the neoliberal culture of individualized professionals that oriented the community into a network of market exchange.

The community was composed of two loosely organized groups, each of them assuming complementary roles in nurturing a movement consciousness anchored in the promotion of community-based, non-profit, permaculture experiments. The Permaculture Institute of Eastern Ontario (PIEO)—now operating under the name Living hEarth—was officially launched in 2011 during the first Toronto Permaculture Convergence. With increasing interest in permaculture in the region, PIEO was in charge of formal instruction through the delivery of PDCs, which were not offered on a permanent basis anywhere in Ontario at the time. The three original founders were land-owning permaculture practitioners living in eastern Ontario. They all completed permaculture trainings in Australia, California, and France, one of them under the instruction of Bill Mollison (PIEO personal correspondence June 3, 2012). PIEO delivered its first PDC in the fall of 2011 and has since offered PDCs and introductory workshops every year (PIEO website 2012). PIEO was not a registered business. The three instructors organized as an informal partnership of independent contractors without relationship to any other permaculture institute. However, the Permaculture Research Institute (PRI) recognized the courses they deliver on their Global Permaculture site (PIEO personal communications, December 14, 2012).[4]

Permaculture Ottawa (PO) was the other component of the eastern Ontario permaculture community. It was also founded in 2011 to deliver popular education and to support community-based permaculture design projects. This rather small, not-for-profit community ranged from between 4 and 32 participants during its general assemblies, with a core group of 14 organizers. Nonetheless, approximately 800 members participated informally in meetings, events, and through its online Facebook group (PO Facebook Group 2012). Most new members joined the movement out of a general interest in permaculture as a method of growing food sustainably in urban areas. PO operated as a working group of the Ottawa chapter of the Ontario Public Interest Research Group (OPIRG 2012), a provincial network of university-based, social and environmental justice-focused, incorporated non-profit groups. The meetings and events of PO were inspired by anarchist insights promoting participatory committees and assemblies, consensus-based decision-making, as well we anti-oppression practices (PO meeting minutes, Nov. 11, 2011). PO's organizers came from various socio-cultural backgrounds; most of them from white, middle-class, urban communities in Ottawa. Four of the core members were self-employed as contractors or business owners, three were employed as university students, two were still employed or transitioning out of employment with the government as public servants, two were retired, one was a civil society professional, one was a teacher, and one person was receiving support while living with a disability. Only one person among the core organizers was a person of color. Eight out of the fourteen organizers were women. All organizers identified as cisgender and heterosexual. Two organizers were recent migrants to Canada.

A prevailing sentiment among the leadership was a desire to leave their current, precarious, or alienating occupation/work to eventually engage in and live from direct, land-based production (farming, bee keeping, herbalism, etc.). In this respect, the community resembled the downward class mobility of the "back to the land" movement of the 1960s and 1970s (Jacob 1997). The PO leadership attempted to orient its actions in solidarity with social and environmental justice struggles in Canada and the United States, following in that sense the popular education model of "liberation permaculture". The latter concept was first coined at a 2007 PDC organized by the Oakland-based Movement Generation in California. Movement Generation mobilizes in opposition to capitalism, racism, and colonialism, through permaculture projects with impoverished and racialized communities.

[4] PRI's Global Permaculture site allows local permaculture institutions, teachers, and students to create their own profile, which the PRI then verifies to ensure that teachers and students did in fact complete a recognized PDC.

> Our vision for fostering a multi-sectoral and holistic movement for ecological justice is rooted in deliberate and coordinated mass action among key racial, economic & environmental justice organizations in the Bay. Our approach in bringing Permaculture for the People is to make it a step towards this end (Movement Generation 2013).

In the summer of 2012, PO held several lengthy brainstorming sessions on the direction and identity of the group, one of them in collaboration with PIEO to envision the possibility of a common basis of unity. After a few months of reflection and discussions, "liberation permaculture" was rejected as a common basis since numerous members from both PO and PIEO expressed concerns that this label was too radical and would exclude some people who did not share such political views (field notes and PO meeting minutes, August 22, 2012). Instead, PO tacitly decided to leave the political objectives at the heart of their practices undefined in order to focus on organizing community activities and projects. Without an explicit political position, everyone could decide to join or remain in PO based on a self-identification as permaculture practitioner. Sustained participation with PO therefore depended either upon the completion of a PDC or similar experiences, or the desire to assume administrative roles (field notes based on participation with PO). Membership in the PO community became oriented around practitioner "technical" knowledge and expertise rather than its initial commitment to an emancipatory political vision.

The PDC offered by the PIEO began to function as the focal point around which members of the permaculture community—especially among core organizers—came to know one another. PDCs therefore became the predominant symbolic gateway for long-term and committed participation among PO members, demonstrating a tacit commodity relationship acting as a social node for the eastern Ontario community, despite great efforts by PIEO to avoid such commodification. For instance and unlike convergences elsewhere, the second Permaculture Convergence of eastern Ontario held in 2012 did not require a PDC or previous knowledge of permaculture to attend any session as organizers were seeking to facilitate broad participation. Yet, membership quickly became primarily composed of those who had completed a PDC, which put pressure on non-PDC practitioners to take the course in order to be fully recognized within the community (field notes based on participation with PO). Hence, PO was inadvertently guiding people towards paying PIEO to obtain their "expertise" in permaculture thereby participating in the market-based means of legitimation by the global permaculture network.

Permaculture Ottawa no longer exists as an organization but a community of the same individuals practicing permaculture as a component of their livelihood persists. By necessity, the eastern Ontario community dissolved because organizers became too busy pursuing permaculture initiatives in a diffused, individual, and private manner. This tendency resembles the initial vision of Mollison and Holmgren for individuals' withdrawal from industrial society rather than the social justice vision articulated by Movement Generation through liberation permaculture as a collective political project. Despite the initial aspirations among the community's leaders to become a movement for liberation permaculture, the permaculture community of eastern Ontario ended up being constrained mostly by individual economic needs. Here, it is important to note that the racialized and working-class context of Movement Generation's liberation permaculture emerged out of the requirement for organized political resistance to ensure survival. In contrast, the material pressures on the eastern Ontario permaculture trainers to sustain themselves were articulated by members of the community through the typically white and middle-class means of individualized entrepreneurship involving privately-owned land and/or enterprises.

The PO experience suggests that communities aspiring towards post-capitalist alternatives within a white, middle-class context are prone to fall back upon reproducing market-based social relations, because of their relatively privileged position within the neoliberal capitalist order. This privilege affords them the ability to conform to market relations rather than having to engage in political resistance like those who are excluded from neoliberal capitalism. In the case of the eastern Ontario permaculture community, practitioners attempted to escape the industrial model of agricultural production; yet the lack of a communal land or collective space and the institutional structure of the PDC directed its members towards individual entrepreneurialism, embedded in market relations, rather than fostering a political strategy of resistance. Based on Gramsci's theoretical insights, the following section will highlight some of the successes of the agroecological pedagogy used by the MST and the ELAA schools to escape the systemic pressure towards reproducing market relations within an agrarian movement. By doing so, we will draw useful lessons to potentially rethink and transform the pedagogy and practices of permaculture design and activism in Canada and advanced industrial regions more generally.

Pedagogies of political agroecology among marginalized communities in Brazil

> We must participate in all popular demonstrations; in the religious festivals of each state, in carnival, in elections, and in mass struggles. It is there that the people are, it is there that the popular consciousness is (Ademar Bogo, MST leader, cited in Karriem 2013, p. 142).

Contrary to the lack of research on permaculture, there is a growing number of studies on agroecology, as well as many peasant organizations seeking to implement such practices, especially in Central and Latin America (Altieri and Nicholls 2017; Levidow et al. 2014; Meek 2015; Meek and Tarlau 2016; Toledo and Barrera-Bassols 2017). Agroecology takes into account the specificities of each farming context and it promotes agricultural practices that are respectful of ecosystemic cycles, contributing to building soil fertility and to reducing the use of agrochemicals and energy inputs (Altieri 1995; Pimbert 2008). Similar to permaculture communities, agroecology advocates do not agree whether it should involve a politicization process to contest the dominant model of development and agriculture, and to promote alternative worldviews and broader social change. A recent divide has in fact emerged within the research community between: (1) researchers seeking to solve existing problems within the dominant, neoliberal food regime by introducing and sharing agroecological techniques, as well as mitigating negative environmental impacts; and (2) activist-researchers committed to, and/or working with, peasant and farmer movements. In the latter case, we use the concept of "political agroecology" to refer to those who are promoting a critical pedagogy that includes a political agenda which goes beyond the technical aspects of agroecology. They are explicitly advocating for a radical transformation of the neoliberal and colonial agri-food regime and they presently seem to dominate academic debates, at least based on a cursory literature review (Altieri and Nicholls 2017; Meek and Tarlau 2016; Mendéz et al. 2013; Hadich and Tardín 2009).

Within a political agroecology framework, scholars and activists are adopting the role of organic intellectuals insisting on the necessity to rethink our modes of living and being in the world (Méndez et al. 2013; Taylor 2015). Their analyses highlight the pervasive and long-term consequences of the Western-led, consumerist, and individualistic society, while pointing toward more sustainable and just practices. Their political agenda defies today's structures of power and governance, as well as the dominant norms and values that are shaping education systems and everyday lives in the global north and south (Meek 2015; Meek and Tarlau 2016). Hence a Gramscian approach is useful to highlight how agroecology is a popular education project to foster social change; it calls for a "war of position" in order to build counter-hegemonic movements. These processes are posited as essential struggles to open greater spaces for imagining alternative models of agriculture, development, education, and modes of living, to be nurtured and consolidated alongside the dominant structures of power and governance.

The historico-political context and the centrality of the pedagogical approach of the Brazilian Landless Workers' Movement (MST) are both essential to understand their struggle. Land in Brazil has been an exceptionally contentious issue, like in many Latin American countries. Land also remains a key factor explaining the extreme inequality among Brazilians. From its early days in the 1980s, the MST leadership has proposed a broad politico-territorial project to bring about social change through democratic and popular struggles, in response to the increasing opposition to the military regime (Stedile and Fernandes 1999; Karriem 2013). Facing gross inequalities and oppression and building on the previous experiences of the peasant unions, the MST was explicitly calling for a socialization of the means of production. Massive land occupations became the landless movement's main tactic for pressuring the federal government to proceed with agrarian reform. Indeed, during their first National Congress in 1985, they claimed that "Without land reform there is no democracy."[5] The legitimacy of occupations is based on the 1988 Brazilian Constitution, which requires the redistribution of unproductive lands for social purposes to those who want to cultivate it.

Participants in land occupations have a lot to gain, including access to a plot, means of subsistence, education, and dignity. Nonetheless, as analysts have demonstrated, life in encampments remains risky, without guarantee of success and with numerous hardships, violence, and criminalization (Branford and Rocha 2002; Stedile and Fernandes 1999; Massicotte's field notes). After a number of isolated occupations by landless peoples across the country in the late 1970s and early 1980s, the idea of uniting forces emerged. The MST was officially created in 1984 as a national movement to coordinate occupations and to mobilize greater number of people and resources, thus reducing the risks of repression and eviction. After more than 30 years of struggle, the MST is therefore a good example of organic intellectuals emerging through direct actions and participation among marginalized sectors. Since the beginning, its leaders insist on the importance of mass-mobilization, formation, and sustained participation to strengthening the capacity of (landless) peasants as part of the Brazilian working class. In an interview during the National Festival of Art and Culture for Agrarian Reform in Belo Horizonte, for example, a leader of the MST, João Pedro Stedile, noted:

> … the way out of these [economic, political, environmental and social] crises that Brazil experiences today depend on the construction of a new project and vision, that can unite a majority. There is a need to build a new vision. … [knowledge] is the main weapon that we have if we want to change the world we live in. The working class needs to understand capitalism if it

[5] MST website, accessed December 10, 2017: http://www.mstbrazil.org/content/history-mst.

> wants to transform it. ... The arts and culture, including culinary art, can and should be utilized as methods that cultivate alternatives to capitalism in the working class (Stedile in Bernardes 2016, no page).

Like Gramsci, Stedile recognizes the importance of knowledge and the cultural sphere to build coalitions among marginalized sectors and allies to develop a "popular" political project for transforming contemporary societies.

The crucial role of the MST's allies should also be noted, in particular that of the Comissão Pastoral da Terra (Pastoral Land Commission, or CPT), which was created in 1975 by advocates of liberation theology.[6] During the early years of the movement, the CPT played a key role in supporting the demands of the poor and accompanying MST participants in their search for dignity, greater autonomy and control over their lives. The MST also has been working with progressive rural and urban forces, from church-based local communities, organized workers and students to transnational networks such as Vía Campesina and Climate Justice Now! These alliances also helped them to put pressure on political authorities at various scales and to secure significant gains, including collective access to land and credit. The direct action and mobilizing tactics of the MST did bear fruits since it became one of the most respected, well known, and largest grassroots movements across the Americas.

It was in the mid-1990s that the MST began to integrate more seriously the ecological aspect of agrarian reform and its centrality in building a different society. Indeed, the first *assentamentos*, or settlements, emerged at the end of the 1980s, when the Green Revolution principles were appealing among farmers. These principles included the use of high-yield varieties, heavy machinery, irrigation, and chemical fertilizers, which are capital and energy-intensive techniques. Many MST settlers were dreaming of high productivity through the use of big tractors and pesticides that were perceived as the best way to improve people's life and food security. This model of farming however relied on access to credit and growing indebtedness. The alarm bell sounded when family farmers became increasingly dependent on financial institutions and when the use of agrochemical inputs resulted in soil and water contamination, as well as health problems in their communities. As neoliberal advocates failed to fulfill their promises, a growing number of voices within and outside of the MST questioned the intensive model of industrial monocultures, calling for just and ecological practices, oriented toward the well-being of communities and embedded in harmonious interactions with the environment.

Agroecology thus became, after much debates, a central platform of the MST struggle. It is now a key part of its political agenda to fight capitalism and the Brazilian policies on industrial agriculture, international trade, and development (Pahnke 2015). As a pedagogy, agroecology was strategically integrated into every training and curriculum of the movement, from kindergarten to adult literacy courses, teachers, political and leadership trainings, as well as technical and university degrees focusing on agricultural production (interviews by Massicotte, 2008, 2009). Nevertheless, one should note that when MST participants first occupy land, as in many economically poor areas, they rely on agroecological techniques for lack of access to mechanical tools or agrochemical products, as a survival strategy. The physical territory of encampments thus represents a living school where everyone needs to participate, thus learning how to organize collectively and to ensure everyone's security, in part through agroecological practices. As the MST argues: the encampment itself is a giant school. Whoever spends time there will never be the same again, since the direct action that takes place there, in and of itself has a formative and educational nature (quoted in Barbosa 2017, p. 126).

As part of their commitment to agroecological pedagogy, the MST became the main promotor and coordinator of the Latin American School of Agroecology (ELAA) that was launched in August 2005 with its first cohort of about 70 students. The school is located inside the Contestado settlement of the MST, in the state of Paraná, Brazil, but it is part of the transnational network Vía Campesina. Indeed, during the 2005 World Social Forum held in Porto Alegre, Brazil, an agreement was signed between peasant movements, the Federal University of Paraná, and governmental institutions, including the governments of Venezuela and of the State of Paraná, who all agreed to support the project (Massicotte 2014). Like the MST, the ELAA has developed a unique pedagogical program that is grounded in and building on the daily life and knowledge of poor rural workers. Indeed, the school is attended by mostly young, subsistence peasants and family farmers who are active within a rural organization. Most of them are from Brazil and the Southern Cone countries, especially since new agroecology schools have been created for other regions of the continent. Overall, the university training is 3-and-a-half-years long, alternating between periods of 60–75 days of training at the ELAA school and 90 days in their home community. Both components of the program are integral parts of the curriculum in order for the student-farmers to share and put the acquired knowledges into practice while they go back to their communities. In this sense, each "student" is considered to be an "activist-technician-teacher" of agroecology, responsible

[6] Paulo Freire and his well-known book the *Pedagogy of the oppressed*, was a popular educator who greatly inspired the MST pedagogy and liberation theology. Considering our main focus here on Gramscian insights, we decided not to include a discussion of Freire's important influence on the MST, but see for example Barbosa (2017).

for sharing such knowledges (Hadich and Tardín 2009, p. 4; Massicotte 2014).

Similarly, Pahnke observed during his 2011 field research at the *Escola Milton Santos* in Paraná that the MST was "encouraging localized research" and that agroecological "technicians were learning how to convince member families to decommodify their production practices by learning how to do without purchased inputs and thus autonomously reproduce their means of existence" (Pahnke 2015, p. 1098). Altieri and Toledo (2011) also insist that techniques in agroecology are mostly knowledge intensive, as opposed to input intensive; they are adapted to each site and have to build on peasants' knowledges and know-how. Hence, we can see how the pedagogy that both the ELAA and MST are actively supporting the constant training of members, starting from their knowledge and experiences, as a key element to improve their own lives and movements, but also to bring about significant change on a much broader scale. Following insights from a Gramscian approach, agroecological education is part and parcel of the emancipatory political project put forward by organic intellectuals emerging from within these movements. As Barbosa aptly pointed out (2017, p. 119), what the MST refers to as "educação do campo", or education by and for the countryside, is central to build a "critical awareness" among marginalized rural sectors and forge a "powerful historical-political actor" struggling for social change that responds to their own needs.

While at the ELAA, participants divide their time between theory and practice, where everyone needs to complete homework and take part in interactive discussions of technical as well as philosophical, political, and economic issues related to agroecology, social change, and movement-building (Massicotte's field notes). Moreover, they have practical, hands-on learning in the field to experiment with pest control or composting, for example. The training is also organized in such a way that everyone participates in the daily functioning of the school. On a typical day, based on a rotating system, a group of students is in charge of cooking, while another one is organizing leisure time or cleaning common areas (bathrooms, common eating space, classrooms), thus learning to build community. At the ELAA, borrowing directly from MST routines, each day begins with the *mística* led by a group of apprentices. This is a collective ritual through which they celebrate the accomplishments of their movements, important moments and figures/heroes, or thank the earth for sharing its riches and beauty. The *mística* is therefore part of the training and it contributes to strengthen the identity and sense of belonging to peasant organizations and the Vía Campesina. The coordinators of the school also insist on nurturing solidarity, democratic decision-making, autonomy, and mutual learning (Massicotte 2014; field notes and interviews), depending on everyone's skills and ideas. Some participants teach their native language or music, others help to paint, post information on the website, share their knowledge of medicinal plants, or fix a leaking toilet. In doing so, they learn to live as a collective and to count on one another. The solidarity element and cooperation are also critical as the financial resources of the ELAA are quite limited. Hence, the School still relies mostly on voluntary trainers, activist-scholars and agroecologists to come on their own time to share their knowledge and experience, and to accompany the activist-technician-teachers in their undertaking.

This reliance on individuals' willingness represents a fragility of the ELAA since there is no financial security, nor a permanent team of trainers guaranteeing the continuity of the program. During Massicotte's last visit to the School for example, in 2010, there were 4 full time staff to coordinate the program, which included two specialists of agroecology who participated on a regular basis in the training activities. They were both members of the MST, like other staff persons, who helped with logistics and maintenance of infrastructure. In general, these are deeply committed activists who do not receive salaries, but their subsistence is covered by the MST and/or Vía Campesina. Furthermore, even if students only need to pay for transportation costs between ELAA and their community—which is often covered by their farming organization—many are not completing the full curriculum for lack of financial resources. The funding of governmental programs supporting the agroecology school was also essential to allow this experiment to take roots, but it remains greatly dependent on the capacity of the MST to maintain its pressure on political authorities (Barbosa 2017). Nonetheless, solidarity has been central to the completion of at least three cohorts of students by relying on the help of allies and voluntary work from various brigades—non-remunerated workers from the MST and allied organizations—as well as the School's physical site and the construction of basic infrastructures (Massicotte, field notes). As the coordinator of the ELAA explains, agroecology "integrates the cultural transformation of human beings, to strengthen the organizations that struggle for transforming our society" (ELAA interview, author's translation). Hence, Tardín highlights the intentional efforts of agroecology leaders to encourage every participant to think critically about their cultural practices and to be active citizens struggling for just and sustainable societies.

While visiting the ELAA school in Paraná, every MST leader, student, activist and trainer that we discussed with insisted on the transformative potential of agroecology, in line with the definition provided by a leading agroecology researcher, Gliessman (2013; Méndez et al. 2013). Gliessman emphasizes agricultural "knowledge and methodology" that go "beyond technical training" and allow participatory decision-making by farmers themselves, based on their "local and empirical knowledges" of ecosystems. Hence, a core element of agroecological pedagogy and the transition process is a commitment to movement-building through mutual learning, participation and everyday practices. Another key example of the application of such praxis is through what Brazilians call the *Diálogo de saberes*, or dialogue of wisdoms, that is a form of mutual learning between scholars and peasants, in which both are contributing valuable knowledges and neither claims a full understanding of everything (Hadich and Tardín 2009). This pedagogical process is trying to break with the often-mutual disdain between both groups by building trust through respectful discussion and exchanges of ideas, practices, beliefs, histories, and hopes. This type of interactive learning is actively cultivating mutual understanding and sharing of experiences that are more holistic and inclusive of the multiple forms of knowledges and various ways of living and working on the land.

Conclusion

This paper invites permaculture communities to begin a conversation about what the praxis and pedagogy of PDCs could resemble if they seek to decommodify instruction and engage with other movements in their struggles for innovations and social change. Peasants' organizations across the world have documented the political gains made through the use of critically engaged pedagogies within movements, particularly among groups advancing a vision of agroecology (Motta 2017; Barbosa 2017; Wright 2014; Ploeg 2009). Some governments, including those of Bolivia and Ecuador, have legally recognized and adopted food sovereignty and agroecology as part of their political agenda. Hence, there is a lot that permaculture communities can learn from political agroecology movements, as well as from the two pedagogical praxes that were presented in this article.

The first lesson permaculture communities can learn from the ELAA and the MST experiences is the need for permaculture training, and especially PDCs, to be developed within the broader counter-hegemonic contexts of Canada's most oppressed populations rather than within a white, middle class setting. First, instructors should take an explicitly political approach to permaculture education towards social change; one that is confronting systems of neoliberal capitalism and colonialism and which is respectful of the diversity of experiences (Altieri and Nicholls 2017; Meek and Tarlau 2016; Mendéz et al. 2013; Hadich and Tardín 2009). It is also important to recognize the potential place of white, middle-class permaculture communities within a broader counter-hegemonic movement seeking to displace capitalism and colonialism in Canada. Permaculture training in this sense should not be forced upon oppressed communities; rather, it should take direction from their actions opposing industrial agriculture and transform the concept of permaculture itself to serve a broader counter-hegemonic praxis. The value of this approach is evident when considering the success of the MST and Movement Generation's liberation permaculture, which arose out of racialized and/or poor social contexts as radical visions challenging and providing alternatives to the corporate food regime. From this insight, PDC instructors could foster collective capacity, solidarity, and movement-building within their communities. By privileging alternatives to industrial agriculture that emerge from oppressed communities, they could emphasize that marginalized communities have ownership over such practices, while adding permaculture design insights when appropriate or asked upon. Permaculture could positively contribute from its own socioeconomic location, to a counter-hegemonic model for food production. However, it needs to engage on a grassroots level with oppressed populations to figure out what a shared vision for just and sustainable food systems would resemble, without expecting or seeking to impose its own vision and/or PDCs as the appropriate or universal "model". This means effectively placing the means of agricultural knowledge production and instruction under the control of oppressed communities.

A related lesson that the article draws from the Brazilian experience is that, despite decades of organizing and a strong network of local, national and transnational alliances with progressive forces, the reliance on volunteer work for the delivery of agroecological instruction was problematic in the ELAA experience. However, the current model of pedagogy in permaculture, despite the best intentions of instructors and participants, reinforces the commodity relationships that compel participants to replicate individualized, market-oriented production in their own permaculture initiatives. The ELAA example, on the other hand, situated voluntary instruction within a post-capitalist solidarity economy to accommodate the material needs of instructors and activists-learners-farmers. An emancipatory form of permaculture training could be established if it were to be led and defined by oppressed social sectors within the solidarity economies of movements, especially those mobilizing against predatory capitalism and settler-states such as Canada and the USA. Ultimately, it is members of oppressed communities themselves whose capacity needs to be valued and supported to lead training as organic intellectuals and to articulate an emancipatory vision of just and sustainable agri-food systems in Canada. Then from such an emancipatory vision of permaculture, white, middle-class practitioners could take direction from, share resources, and act in solidarity with oppressed communities towards common goals.

The second lesson to be learned from the ELAA is the need for much greater horizontal collaboration, mutual learning, and locally grounded experimentation that is respectful of the specificities of both ecosystems and communities, rather than the pre-set design process envisioned by Bill Mollison and taught through PDCs. To be able to sustain a transition toward just and sustainable agri-food systems, many advocates and practitioners of permaculture and political agroecology have to work together toward greater solidarity, collective autonomy, and well-being. Today, small-scale farmers and potential producers in Canada who are seeking farmland are confronted by powerful socio-economic and political forces with little flexibility toward alternative modes of production and social reproduction. Debt-burdened and underemployed new farmers interested in permaculture find themselves in competition with industrial agribusiness and deeply bureaucratized state institutions for land access. These agrarian challenges facing small-scale permaculture practitioners raise questions of how to bring about a system of equitable land distribution under a neoliberal regime and Canada's ongoing system of colonialism. Such structures of exploitation have a much greater impact on the livelihoods of Indigenous peoples. However, they stand as common systems of exploitation that link the experience of oppression of small-scale, ecological farmers and Indigenous peoples alike through an inequitable distribution and overexploitation of land. Permaculture communities in advanced industrial regions should therefore learn from some of the political successes of the MST struggling against the violence of dispossession, de-peasantization, and criminalization of its members by pressuring the state and gaining important legal and policy reforms that gave them a collective control over significant extension of land and rural education (Barbosa 2017). In Canada, this lesson would suggest forming alliances with other groups facing land struggles such as Indigenous peoples, by reimagining what PDCs could resemble if shared by emancipatory forces and existing movements, including the National Farmers Unions or Union Paysanne (Kepkiewicz and Dale 2018). These alliances could reorient permaculture communities away from commodity production, towards a socially transformative practice of agroecological design.

The MST's struggle for control over farmland and collective autonomy has been informed by agroecological praxis and critical pedagogy that include an emancipatory political project calling for the consolidation of a self-conscious counter-hegemonic movement. The MST pursues this project through the expansion of its alliances with progressive urban forces to engage in a more effective war of position within civil society. The mass membership, the insistence on continuous participation and capacity-building of its members, and the numerous allies of the MST contributed to the success of the ELAA. Learning from these, certainly partial and relative but successes nonetheless, and the limitations faced by PO suggest that permaculture communities in Canada will confront material and institutional barriers in their attempts to transform the dominant agri-food system, especially if they continue to organize as mostly white, middle-class people independently of larger social movements. Based on our research and review of the literature, we argue that similar to advocates of political agroecology, Canadian permaculture communities could improve their chance of promoting a transition towards alternative agri-food production, social reproduction, and equitable land distribution by working in broader alliances with Indigenous peoples, women, and other oppressed groups. They could accomplish this by collaborating in the formation of a decommodified, critical, and politicized vision of permaculture instruction in support of these movements.

Recent literature (Altieri and Toledo 2011; Barbosa 2017) has provided a better understanding of the crucial role of critical pedagogies and organic intellectuals, such as agroecology and permaculture instructors, in mobilizing alternative visions of agri-food production within social movements. By actively promoting alternative norms, emancipatory pedagogies, and practices, embedded in the lived experiences of marginalized groups, these actors play a crucial role in consolidating grassroots movements that are defying the dominant common sense surrounding both the politicization of permaculture, and ecological and social justice in food production and social reproduction. By openly reflecting on and exposing the socioenvironmental impacts of the current corporate agri-food system, permaculture training would politicize food and agricultural concerns, and hence contribute to mobilizing a greater number of individuals and communities in the struggles for alternatives to industrial agriculture. We hope that this analysis will help to sow the seeds of a much-needed reflection on the benefits and limitations of different forms of pedagogical praxis in Canada and beyond.

Acknowledgements Our gratitude goes out to all interviewees and activists, especially those from the Ottawa permaculture community, the ELAA and the Brazilian MST for sharing their knowledge and expertise informing this research. Your love for your community and dream for a sustainable world is inspiring. Thanks to Patricia Ballamingie and Jill Wigle for supervising the eastern Ontario fieldwork that went into the writing of this paper, and to reviewers for providing insightful feedback to improve this article. A special thanks to Tiaraju P. D'Andrea for research assistantship, as well as Stephen Brown, Peter Andrée, Alicia Martin, and Michael J. Wigginton for offering their feedback and editing. This paper was produced with the support of the Social Sciences and Humanities Research Council, the Ontario Graduate Scholarship, and The Douglas Fullerton Award in Urban Studies.

Compliance with ethical standards

Ethical approval All procedures performed in studies involving human participants were in accordance with the ethical standards of the institutional and/or national research committee and with the 1964 Helsinki declaration and its later amendments or comparable ethical standards.

References

Aiken, G. T. 2017. Permaculture and the social design of nature. *Geografiska Annaler: Series B, Human Geography* 99 (2): 172–191. https://doi.org/10.1080/04353684.2017.1315906.

Akram-Lodhi, H. 2013. *Hungry for change: Farmers, food justice and the agrarian question*. Winnipeg: Fernwood.

Altieri, M. A. 1995. *Agroecology: The science of sustainable agriculture*. Boulder: Westview Press.

Altieri, M. A., and I. C. Nicholls. 2017. Agroecology: A brief account of its origins and currents of thought in Latin America. *Agroecology and Sustainable Food Systems* 41 (3–4): 231–237. https://doi.org/10.1080/21683565.2017.1287147.

Altieri, M., and V. Toledo. 2011. The agroecological revolution in Latin America: Rescuing nature, ensuring food sovereignty and empowering peasants. *The Journal of Peasant Studies* 38: 587–612. https://doi.org/10.1080/03066150.2011.582947.

Bang, J. M. 2005. *Ecovillages: A practical guide to sustainable communities*. New Society Publishers.

Barbosa, L. P. 2017. Educação do campo [education for and by the countryside] as a political project in the context of the struggle for land in Brazil. *The Journal of Peasant Studies* 44 (1): 118–143. https://doi.org/10.1080/03066150.2015.1119120.

Bernardes, J. 2016. *Culture is the best way to build working class unity*. Friends of the MST website, 21 June, access online, April 20th, 2017: http://www.mstbrazil.org/news/culture-best-way-build-working-class-unity.

Bookchin, M. 1962. *Our synthetic environment*. New York: Knopf. (Published using pseudonym Lewis Herber).

Branford, S., and J. Rocha. 2002. *Cutting the wire: The story of the landless movement in Brazil*. London: Latin America Bureau.

Brawner, J. 2015. Permaculture in the margins: Realizing Central European regeneration. *Journal of Political Ecology* 22: 429–444. https://doi.org/10.2458/v22i1.21117.

Carson, R. 1962. *Silent spring*. Boston: Houghton Mifflin.

Conford, P. 2011. The organic alternative: Farming. In *The development of the organic network: linking people and themes, 1945-95*, 71–135. Edinburgh: Floris Books.

Dawson, J. 2006. *Ecovillages: New frontiers for sustainability, Schumacher Briefing*. White River Junction: Chelsea Green Publishing.

Ferguson, R. S., and S. T. Lovell. 2014. Permaculture for agroecology: Design, movement, practice, and worldview. A review. *Agronomy for Sustainable Development* 34 (2): 251–274. https://doi.org/10.1007/s13593-013-0181-6.

Ferguson, R. S., and S. T. Lovell. 2015. Grassroots engagement with transition to sustainability: Diversity and modes of participation in the international permaculture movement. *Ecology and Society*. https://doi.org/10.5751/ES-08048-200439.

Ferguson, R. S., and S. T. Lovell. 2017a. Diversification and labor productivity on US permaculture farms. *Renewable Agriculture and Food Systems*. https://doi.org/10.1017/S1742170517000497.

Ferguson, R. S., and S. T. Lovell. 2017b. Livelihoods and production diversity on US permaculture farms. *Agroecology and Sustainable Food Systems* 41 (6): 588–613. https://doi.org/10.1080/21683565.2017.1320349.

Fukuoka, M. 1990. *The one-straw revolution: An introduction to natural farming*. (Trans. L. Korn) New York: NYRB Classics.

Gliessman, S. 2013. Agroecology: Growing the roots of resistance. *Agroecology and Sustainable Food Systems* 37 (1): 19–31. https://doi.org/10.1080/10440046.2012.736927.

Gordon, U. 2008. Luddites, hackers and gardeners: Anarchism and the politics of technology. In *Anarchy Alive! Anti-authoritarian politics from practice to theory*, 109–138. London: Pluto Press.

Gramsci, A., and Q. Hoare. 1971. *Selections from the prison notebooks of Antonio Gramsci*. New York: International Publishers.

Hadich, C., and J. M. Tardín. 2009. *Escola Latino Americana de Agroecología: Experiências camponesas de agroecología*. Paraná.

Holmgren, D. 2002. *Permaculture: Principles and pathways beyond sustainability*. Hepburn: Holmgren Design Services.

Holmgren, D. 2011. The long view. In *Permaculture pioneers: Stories from the new frontier*, eds. K. Dawborn, and C. Smith, 18–29. Australia: Melliodora Publishing.

Hopkins, R. 2008. *The transition handbook: From oil dependency to local resilience*. White River Junction: Chelsea Green Publishing.

Jacob, J. 1997. *New pioneers: The back-to-the-land movement and the search for a sustainable future*. Pennsylvania: The Pennsylvania State University.

Jungck, J. R. 1985. Perennial polyculture, permaculture and preservation: The principle of diversity. *The American Biology Teacher* 47 (2): 72–75.

Karriem, A. 2013. Space, ecology, and politics in the praxis of the Brazilian landless movement. In *Gramsci: Space, nature, politics*, ed. Michael Ekers, Gillian Hart, Stephen Kipfer and Alex Loftus, 142–160. Oxford: Wiley.

Kelly-Bisson, C. 2013. *Forests for the people: Resisting neoliberalism through permaculture design*. Department of Geography & Environmental Studies, Carleton University.

Kennedy, D. 1991. Permaculture and the sustainable city. *Ekistics* 58 (348/349): 210–215.

Kepkiewicz, L., and B. Dale. 2018. Keeping 'our' land: Property, agriculture and tensions between Indigenous and settler visions of food sovereignty in Canada. *The Journal of Peasant Studies*. https://doi.org/10.1080/03066150.2018.1439929.

Lawton, G. 2011. Thinking big. In *Permaculture pioneers: Stories from the new frontier*, ed. K. Dawborn, and C. Smith, 87–109.

Levidow, L., M. Pimbert, and G. Vanloqueren. 2014. Agroecological research: Conforming—or transforming the dominant agro-food regime? *Agroecology and Sustainable Food Systems* 38: 1127–1155. https://doi.org/10.1080/21683565.2014.951459.

Mann, S. 2012. September 27. Episode 1223: *The cost of a permaculture design course (Permabyte). The Permaculture Podcast*. Retrieved from http://www.thepermaculturepodcast.com/2012/pdc-cost/.

Massicotte, M. J. 2014. Beyond political economy: Political ecology and La Vía Campesina's struggle for food sovereignty through the experience of the Escola Latinoamericana de Agroecologia (ELAA), Brazil. In *Globalization and food sovereignty*, eds. P. Andrée, J. Ayres, M. Bosia, and M.-J. Massicotte, 255–287. Toronto: University of Toronto Press.

McMichael, P. 2009. A food regime genealogy. *The Journal of Peasant Studies* 36 (1): 139–169. https://doi.org/10.1080/03066150902820354.

McWilliams, C. 1939. *Factories in the field: The story of migratory labor in the United States*. Boston: Little, Brown and Company.

Meek, D. 2015. Learning as territoriality: The political ecology of education in the Brazilian landless workers' movement. *The Journal of Peasant Studies* 42: 1179–1200. https://doi.org/10.1080/03066150.2014.978299.

Meek, D., and R. Tarlau. 2016. Critical food systems education (CFSE): Educating for food sovereignty. *Agroecology and Sustainable Food Systems* 40: 237–260. https://doi.org/10.1080/21683565.2015.1130764.

Méndez, V. E., C. M. Bacon, and R. Cohen. 2013. Agroecology as a transdisciplinary, participatory, and action-oriented approach. *Agroecology and Sustainable Food Systems* 37 (1): 3–18. https://doi.org/10.1080/10440046.2012.736926.

Mollison, B. 1988. *Permaculture: A designers manual*. 2nd ed. Tyalgum: Tagari.

Mollison, B., and D. Holmgren. 1978. *Permaculture one: A perennial agriculture for human settlements*. Tyalgum: Tagari.

Morel K., C. Guégan, and F. Léger. 2016. Can an organic market garden based on holistic thinking be viable without motorization? The case of a permaculture farm. *Acta Horticulturae*. https://doi.org/10.17660/ActaHortic.2016.1137.47.

Morton, A. D. 2007. Global capitalism and the peasantry in Mexico: The recomposition of class struggle. *The Journal of Peasant Studies* 34 (3–4): 441–473. https://doi.org/10.1080/03066150701806257.

Motta, S. C. 2017. Emancipation in Latin America: On the pedagogical turn. *Bulletin of Latin American Research* 36: 5–20. https://doi.org/10.1111/blar.12526.

Movement Generation. 2013. *Mission*. http://www.movementgeneration.org/about-us/mission-and-history. Accessed 13 April 2013

Odum, H. T. 1971. *Environment, power, and society*. 1st ed. New York: Wiley.

Ontario Public Interest Research Group (OPIRG). 2012. Working Groups. Retrieved from http://www.opirgcarleton.org/.

Pahnke, A. 2015. Education of the countryside at a crossroads: Rural social movements and national policy reform in Brazil. *The Journal of Peasant Studies* 42 (6): 1087–1107. https://doi.org/10.1080/03066150.2014.990444.

Permaculture Global. 2017. *Worldwide permaculture projects*. https://permacultureglobal.org/. Accessed 23 Mar 2017.

Permaculture Institute of Eastern Ontario. 2012. *Home Page*. http://eonpermaculture.ca. Accessed 13 Dec 2012

Permaculture Ottawa. 2012. *Permaculture Ottawa Facebook Group*. Retrieved from https://www.facebook.com/groups/PermacultureInOttawa/.

Pimbert, M. 2008. *Towards food sovereignty: reclaiming autonomous food systems*. London: International Institute for Environment and Development (IIED).

Puig de la Bellacasa, M. 2010. Ethical doings in nature cultures. *Ethics, Place & Environment* 13 (2): 151–169. https://doi.org/10.1080/13668791003778834.

Smith, J. S. 1950. *Tree crops: A permanent agriculture*. Washington, DC: Island Press.

Smith, C. 2011. Introduction, ix-xx. In *Permaculture pioneers: Stories from the new frontier*, ed. K. Dawborn, and C. Smith, Tylagum: Melliodora Publishing.

Stedile, J. P., and B. F. Mançano. 1999. *Brava gente: A trajetória do MST e a luta pela terra no Brasil*. São Paulo: Editora Fundação Perseu Abramo.

Taylor, M. 2015. *The political ecology of climate change adaptation: Livelihoods, agrarian change and the conflicts of development*. New York: Routledge.

Toledo, V. M., and N. Barrera-Bassols. 2017. Political agroecology in Mexico: A path toward sustainability. *Sustainability* 9 (2): 268. https://doi.org/10.3390/su9020268.

Van der Ploeg, J. D. 2009. *The new peasantries: Struggles for autonomy and sustainability in an era of empire and globalization*. London: Earthscan.

Vergara-Camus, L. 2014. *Land and freedom: The MST, the Zapatistas and peasant alternatives to neoliberalism*. London: Zed Books.

Veteto, J. R., and L. Joshua. 2008. Environmental anthropology engaging permaculture: Moving theory and practice toward sustainability. *Culture & Agriculture* 30 (1–2): 47–58. https://doi.org/10.1111/j.1556-486X.2008.00007.x.

Wright, S. 2014. Food sovereignty in practice: A study of farmer-led agriculture in the Philippines. In *Globalization and food sovereignty: Global and local change in the new politics of food*, ed. Peter Andrée et al., 199–227. Toronto, University of Toronto Press.

Marie-Josée Massicotte is an associate professor at the School of Political Studies and the Institute for Feminist and Gender Studies at the University of Ottawa, Canada. Her research interests include social movements, social change, alternative economies, agroecology and food politics, with a focus on Latin America and Mexican politics. She is director of the Laboratory for the Interdisciplinary Study of Food (Food Lab), and member of the GRITE—Research Group on the Territories of Extraction. She co-edited a book on *Globalization and Food Sovereignty*, and another one on *L'altermondialisme: forums sociaux, résistances et nouvelle culture politique*. Marie-Josée is presently investigating the strategies and tactics used by women and men from rural and Indigenous communities to defend their territories in the context of extractivist projects, socio-environmental conflicts and contestation. She is interested in questions related to methodologies, critical pedagogy, research action, and knowledge production emerging from activist practices at multiple scales.

Christopher Kelly-Bisson is a PhD candidate in the School of Political Studies and member of the Food Lab at the University of Ottawa. His research explores the phenomenon of financial land grabbing in eastern Ontario and its impact on farmland accessibility to new farmers. He received a Permaculture Design Course training in 2011. Christopher has operated an organic market garden in Ottawa for four years. He conducts research and training for new farmer organizations in the region.

Agriculture and Human Values (2019) 36:595–610
https://doi.org/10.1007/s10460-018-9853-9

SYMPOSIUM/SPECIAL ISSUE

Teaching the territory: agroecological pedagogy and popular movements

Nils McCune[1] · Marlen Sánchez[2,3]

Accepted: 7 February 2018 / Published online: 15 February 2018

Abstract

This contribution traces the parallel development of two distinct approaches to peasant agroecological education: the peasant-to-peasant horizontal method that disseminated across Mesoamerica and the Caribbean beginning in the 1970s, and the political-agroecological training schools of combined consciousness-building and skill-formation that have been at the heart of the educational processes of member organizations of La Via Campesina since the 1990s. Applying a theoretical framework that incorporates territorial struggle, agroecology and popular education, we examine spatial and organizational aspects of each of these models for peasant education and movement-building. Recognizing that the models, their respective contexts, and the dialectical relationships therein have been in constant evolution, we share findings on the movement-place as a territorial system with socio-historical subjectivity, that is, peasant movements as territorially-embedded, collective historical actors. This leads to some conclusions in moving past educational theory that has centered upon individual subjects, and approaching a conception of territory as a subject of learning processes.

Keywords Agroecological education · Peasant-to-peasant · Social movements · Pedagogical mediators

Introduction

Agroecology has 'grown up', since a series of classic texts in the 1980s and 1990s systematically applied ecological principles and measures to agricultural systems, based on extensive empirical research and observation of indigenous traditional land use (Gliessman and Altieri 1982; Gliessman 2007; Altieri 1989; Carroll et al. 1990). The neoliberal model of deregulated capital accumulation, imposed globally with few exceptions since the rise of the United States as a hegemonic world power in the early 1990s, has also produced a world of intellectual and popular resistances for which agroecology is highly relevant. Diverse forms of political expression have included massive protest, national and international boycotts and campaigns, walk-outs, sit-ins, occupations of closed factories, autonomous and self-managed worker cooperatives, and bottom-up processes of agrarian reform (Klein 2007). What has often lacked, however, is a 'popular political project' (Stedile 2002) in the sense of long-term process of accumulation of forces by social sectors opposed to capital's ecocide. In this context, agroecology itself has been 'occupied' by movements and popular struggles, which find in its practice and theory a tool for joining labor and nature to meet human need outside of capital's imperative to exploit (Sevilla-Guzmán and Woodgate 2013).

Such a re-working of agroecology into a more overtly politicized science, practice and movement (Wezel et al. 2009), however, has not been an even process. Some academic institutions maintain an entirely apolitical version of agroecology, implicitly arguing that the application of ecological theory to agriculture is itself a complex science. Other approaches to the field integrate cultural, social, and historical factors (sometimes including patriarchy, racism, war and imperialism) into their analysis of agroecosystems (Ferguson and Morales 2010; Méndez et al. 2013; Meek and Tarleu 2015a). Perhaps more than in the past, there are highly diverse levels of commitment to transformative social change among the various sectors of society that espouse

✉ Nils McCune
nmccune@umich.edu

1 School for Environment and Sustainability, University of Michigan, Ann Arbor, MI, USA

2 La Vía Campesina of Central America, Managua, Nicaragua

3 Rural Workers' Association (ATC) Nicaragua, Managua, Nicaragua

agroecology (Holt-Gimenez and Shattuck 2011; Meek and Tarleu 2015b). As climate variability increasingly puts the profit margin of industrial monoculture at risk, and market-based mechanisms continue to dominate institutional approaches to handling environmental destruction, a new sort of 'agroecology' is appearing on the agenda of international institutions, including 'climate-smart agriculture' and 'sustainable intensification', as options to be folded into conventional agribusiness value chains (La Via Campesina 2015). The purely technical version of agroecology, and its incorporation into the vacuous, dominant 'win–win' discourse, has been denounced as a case of co-optation and domestication of the emancipatory demands of popular sectors (Giraldo and Rosset 2016; Copeland 2018).

An analogous situation is emerging with regard to food system and agroecological education, as more conformist and more critical pedagogies vie for space in educational institutions (Meek and Tarleu 2015b). In the balance hangs the linkages between the academic critique of mainstream food systems, often from positions of class or race privilege, and the movements of people committed to changing food systems, at the same time as they transform class society, gender and race relations. Popular movements recognize a gap between formal agroecological education and agroecological social action, and have increasingly been creating agroecological teaching–learning processes that are permanently linked to, and indeed stem from, political-territorial social mobilization (Fernandes 2017; McCune et al. 2016). Rather than asking—how can we teach political agroecology?—movements tend to ask, what human, political and technical qualities do our cadre need in order to (collectively) take agroecology to scale? This leads to a distinct set of learning objectives and processes.

Today's popular movements argue that learning is not confined to a classroom, but rather can be found in the forms of reflection, collective production of knowledge, and social activation of popular sectors, all of which 'anticipate' the fundaments of the liberated culture and education in a utopian society (Ouviña 2017). In movements linked to territories and which create incipient educational spaces outside the logic of capital, formative processes take place in ways not discussed by conventional educational thought. The particularities of the peasant-to-peasant movement—its horizontal, Freirian basis, its use in promoting sustainable agriculture and agroecology among small farmers, its origins in the Mayan cosmovision and its global adoption by the international peasant movement, La Via Campesina—make for several fronts upon which to base the argument that the PtP movement is anticipating, in the here and now of global capitalism and the crisis of democracy, a cultural content fitting to a distinct historical epoch. In so far as this whole-culture action can be understood pedagogically, it challenges the 'common sense' notion that learning is an individual process that takes place in one human mind at a time. Instead, the PtP movement portends to a collective, territorial learning process that is historically- and place-specific (McCune et al. 2016). This idea is present in the *diálogo de saberes*, an epistemological method used by La Via Campesina (Via) to share, document, discuss, build and mobilize around agroecological knowledge in its 'battle of ideas' with corporate agribusiness and the financial sectors of global capital (Martínez-Torres and Rosset 2014).

A creative dialogue between approaches to agroecological education is underway within Via to get past old and new obstacles in the path of taking agroecology to scale (Rosset 2015). Attempting to bridge social movement education with broad social action around agroecological food systems, the organizations of Via are recovering the Mesoamerican roots of the peasant-to-peasant method of horizontal learning. This stems from the recognition that agroecological learning processes centered upon individuals in the setting of Via's political-agroecology training schools may produce highly competent movement cadre, but does not assure the kind of territorial impact needed for agroecology to change communities and landscapes. Indeed, the more relevant measure of formative processes may not be in the quality of the thinkers they produce, but in the territoriality of the movement they reproduce.

In this contribution, we analyze the historical emergence, roots and relationships of the peasant-to-peasant (PtP) method and political-agroecological training schools, based on a description of their early interactions in Guatemala during the 1970s. The two approaches share common traits: both cite Brazilian pedagogue Paulo Freire as a major inspiration, and both have had a similar radius of action—fundamentally in Latin American peasant organizations, with a handful of African experiences. Additionally, both approaches have increasingly focused on scaling out agroecological practice, from isolated experiences by farmers to landscape-level social action (Rosset et al. 2011; McCune et al. 2014; Muñoz et al. 2014; La Via Campesina 2015).

Remarkably, however, these two peasant-led approaches to agroecological learning have had very little consistent interaction, until the phenomenal success of the PtP movement within the Cuban Via organization, ANAP, was widely publicized (Machín et al. 2010; Rosset et al. 2011). Since then, Via's International Working Group on Sustainable Peasant Agriculture (known as the International Collective on Agroecology, Seeds and Biodiversity since 2013) has promoted debate and exchange among both agroecological traditions, leading to highly productive processes of synthesis and new pedagogical proposals (McCune et al. 2014; Muñoz et al. 2014; McCune et al. 2016). Here, we add to the existing historical narrative of PtP and begin to build a historical narrative of Via's political-agroecological training schools, at the same time as we critically examine the relationships of each approach to the other and to the changing historical context.

Part of the reason for the longstanding estrangement between PtP and more explicitly political social movement educational processes has been that while the former has origins in non-governmental organization (NGO) practice, the latter emerged from the political-ideological education of leftist political organizations and parties. Tensions have existed as NGOs have maintained a strong presence in PtP in ensuing decades, while popular movements have created schools. The weaknesses of each of these actors has been used in discourse by the other: social movements accuse NGOs of creating short-term, dependent projects that disappear when funding runs dry, while NGOs accuse social movements of neglecting local organizing to focus on political advocacy at the national level (Holt-Giménez 2010; Rosset et al. 2011). In the following section, we investigate these tensions in the context of the Guatemalan civil war in the late 1970s. We show that despite the scarce existence of documented interactions, PtP and political education indeed co-existed, interacted and even complemented one another for a period of time in Guatemala.

Next, we elaborate a theoretical basis for education's role in the radical transformation of food and agriculture systems, drawing from the broader context of both PtP and political-agroecological training schools in to strengthen our framework. We argue that peasant movements create 'social situations of development' (Vygotsky 1978) in which the practical life experience itself, more than any lesson plan or method, is actively appropriated by learners as personal growth. Agroecological scaling using either methodology is content-specific and difficult to replicate based on copying or importing the successful techniques of others. Thus, the learning processes that have been achieved in both approaches should be understood in light of the emerging relationship between the collective subjects, the methods and the historical-geographic context. In fact, it is more accurate to consider the movement-territory as a whole system in each historical case of peasant-led agroecological education, thereby opening the door to an analysis of system components, relationships and exchanges.

Finally, we elaborate on our concept of territorial mediators as cultural and organizational elements that bridge critical education and agroecological scaling. While popular education has long recognized the link between learning and transforming social reality, learning theory has been too focused on individual minds and not enough on the potential of creating place-based, social-cultural processes of transformation that include a strong learning component. Territories, encompassing places, living systems and the human culture therein, are explored as educational subjects in the final discussion.

Each one, teach one: the journey and the method

The civil conflicts in Mesoamerica during the latter half of the twentieth century marked the lives of peasant and indigenous farmers, as hundreds of thousands fled their land and territories, either to neighboring countries or to 'the mountain', where they took on the additional identity of clandestine guerrilla organizations. Enmeshed in this complex history are the origins of the peasant-to-peasant (PtP, known in Spanish as *campesino-a-campesino*) horizontal learning method that, with the exile of many of its original Guatemalan Kaqchikel promotors, reached all of Mesoamerica and eventually came ashore in Cuba. There, the PtP agroecological movement transformed the productive systems of one-half the island's small farmers. Cuba's National Association of Small Farmers (ANAP) is a member of Via, so its globally unique experience of massive agroecological scaling using horizontal methods has been documented and widely discussed in rural social movements across the globe in the last few years. Table 1 lists the acronyms used in this contribution.

Guatemala: the spark and the tinderbox

Guatemala in the 1950s was on the brink of broad social change, as centuries of exploitation of Mayan women, peasants and workers by ladino *elites* was finally coming under scrutiny. Nobel prizewinning novelist Miguel Angel Asturias had called attention to the indigenous struggle against displacement in his book, *Hombres de Maíz*, and to decadent state corruption in *Señor Presidente*. Jacobo Arbenz was elected president in 1952 and quickly moved forward a program of agrarian reform that provoked the hatred of the US-owned United Fruit Company. Future Latin American revolutionary Ernesto Guevara was still a vagabond medical doctor until he arrived to Guatemala in 1951 and quickly integrated himself into the circles of progressive young intellectuals supporting the Guatemalan Workers' Party. In the Revolution of '52', the government of Jacobo Arbenz

Table 1 List of acronyms and organizations

Acronym	Organization
ANAP	National Association of Small Farmers (Cuba)
CLOC	Latin American Coordination of Rural Organizations
CUC	Committee of Peasant Unity (Guatemala)
PtP	Peasant-to-Peasant (the method in general)
PCAC	Peasant-to-Peasant Program (Nicaragua)
UNAG	National Union of Farmers and Ranchers (Nicaragua)

created Decree 900, which instituted agrarian reform and local agrarian committees. By 1954, the US Government had seen enough and the CIA orchestrated a *coup d'etat* by which General Castillo Armas took power and swiftly ended land reform. During the ensuing 'agrarian counter-reform' rural elites displaced thousands of indigenous peasant farmers, as the state outlawed and repressed the Arbenz-era agrarian committees.

Having effectively closed the electoral route to social change, the military government and landed oligarchy essentially created the conditions for a leftist insurgency (Valenzuela 2008). After Cuba's successful revolution and land reform, a generation of social reformers began to study the question of armed struggle. While a few clandestine organizations were indeed formed, the major force for social reform was found in the Christian Democratic party, which integrated Mayan Catholic communities into a formidable voting bloc. By the end of the 1960s, surviving members of the Arbenz-era agrarian committees from the municipality of San Martín Jilotepeque, in the central department of Chimaltenango, mentored a group of young people from the community to rebuild local peasant organizations as Peasant Leagues, with the support of national peasant federations. After the first Peasant League was founded in Santa Anita in 1968 and was not subjected to state repression, Peasant Leagues were formed across the municipality to resist processes of displacement of Kaqchikel peasants by *ladinos*. Through their participation in national organizations, the members of the Peasants Leagues of San Martín received courses on leadership, agrarian law and organizing methods.

Throughout the 1960s, the network of village-level Catholic catechists offered a formidable territorial fabric to the Christian Democratic party, which came to be seen as representing many of the reforms demanded by Mayan communities. In this context, Christian Democrat Felipe Alvarez was elected in 1972 and became the first indigenous mayor of San Martín. However, at the national level, the Christian Democratic party was defeated in a fraudulent election in 1974, which eventually led to the erosion of its support base. Meanwhile, autonomous peasant organizing, mingled with coherent support from village- and parish-level structures of the Catholic Church, produced a growing organized sector of the Guatemalan peasantry. Catholic Action was a network of clergy and residents of communities, which dealt with local problems such as water, housing, land, and access to education. The missionary wing of the Church had begun since the 1950s to promote agricultural and credit cooperatives largely made up of indigenous workers. According to Berryman (1984, cited in Althoff 2014: 88) the number of Church-backed cooperative members in Guatemala grew from 27,000 in 1967 to 132,000 in 1976.

In 1972, a group of 78 peasant farmers graduated from a course on 'social promotion' from the Rafael Landivar University, and began a sustainable agriculture program in 14 communities of San Martín, with the support of the US-based NGO World Neighbors. Of this initial group, the vast majority were catechists, active in Catholic Action, and thus already recognized as community leaders. These people were trained in agriculture, health, cooperativism, social work, leadership and self-esteem (Camposeco Cruz 2011). Thus it was the more community-oriented part of the Christian Democratic movement, organized through local churches as Catholic Action, that created the first peasant-to-peasant experiences through a program carried out by an international NGO.

The actual production practices carried out through the method were basically simple, in general having to do with building erosion control works such as horizontal ditches planted with permanent grassy species, using cover crops and green manures to recover soil nitrogen levels, associating crops, and recovering native seeds. Guatemalan promoters created a model of learning-by-doing that would be the major methodological feature passed along to other Latin American contexts. The rule was that 80% of classes take place on the parcel, using hoes, shovels and machetes, while 20% could take place in the shade of the porch and focus on theory. The promoter was expected to lead by their own example, by maintaining the parcel that could guide the planning, design and implementation of sustainable production by nearby farmers. Promoters also participated fully in the agricultural work they recommended on their neighbors' farms.

The notion of farmer promotion in the San Martín version of PtP was linked to the Mayan concept of *chuchubal*, or mutual aid, and to the Catholic notion of service to society. Two fundamental kinds of exchanges were developed: visits by groups of local farmers to the parcel of the promotor, and exchanges of experiences—visits by promoters to other communities to follow up on the trainings they led. Each promoter was responsible for anywhere between ten and twenty farmers who learned on the parcel of the promotor. Parcels in which at least five systems of practice (such as soil conservation, organic fertilization, crop association, distance between plants) were applied by the farmer became known as articulated training centers (*Centros Articulados de Capacitación*).

The earthquake of 1976, which left thousands of human victims in San Martín, was the spark for the Peasant Leagues and the Sustainable Agriculture program of World Neighbors to develop their capacity for coordinated actions. As the military government failed to reconstruct San Martín, local organizations committed themselves to resolving local problems. At the same time, the earthquake alerted large sectors of society to the inept and corrupt national government's unwillingness to represent the interests of the majority of Guatemalans. The diminished Social Democratic party, seen as weak and not representative of a real alternative (in fact, the 1974 Social Democrat candidate who lost the election due to fraud was none other than General Efrain Rios Montt, who took power in a coup in 1982 and was ultimately charged with war crimes and genocide). Across the country, a vast contingent of young people joined clandestine revolutionary organizations that advocated armed struggle as the only alternative for 'refounding' Guatemala based on agrarian, racial and social justice.

Camposeco Cruz (2011) speculates that the original promoters and participants in the San Martín program had little interest in the revolutionary platform. Their position was that the best struggle of the peasantry was the adequate technological management of their farms. However, the deeply felt agrarian injustices were spoken to by the more radical peasant organizations such as the Committee of Peasant Unity (*Comité de Unidad Campesina*—CUC), which had formed in 1972 based on a reflection on the peasant reality in Guatemala. CUC, which means squirrel (*kuuk*) in all Guatemalan Mayan languages, was the representation of a new level of commitment to radical social change in the countryside. People heard of a 'squirrel' that was walking from village to village, waking up the sparrows, bobcats, owls and songbirds, telling all the animals about their rights and the freedom whose time had come (Albizures 2008). After its first national assembly in 1978, the CUC grew into a massive social movement of the politicized peasanty and farm worker sectors. Its platform gave content to the guerrilla struggle, in that it demanded agrarian transformation and respect for the indigenous majority. By the late 1970s, the largest single contingent of participants in the guerrilla struggle of Guatemala came from radicalized Christian base organizations (Althoff 2014), many of them primarily made up of indigenous peasants. The first contacts between leaders of the Peasant Leagues and the Revolutionary Armed Forces (FAR) in San Martín took place in 1978 (García 2012). By the end of 1978, the CUC had declared a 'total commitment to national transformation' and was allied with the largest guerrilla faction, the Guatemalan Army of the Poor (*Ejército Guatemalteco de los Pobres*) (García 2012). By 1980, the CUC led the largest strike in Guatemalan history, as 80,000 sugar cane cutters and cotton pickers stopped working until they gained a substantial pay raise (Comisión para el Esclarecimiento Histórico 1999).

The CUC demanded the ouster of the military government and its replacement by a revolutionary, patriotic, democratic, people's government. To better organize the struggle, the CUC advocated unity among all the 'oppressed people of the countryside': indigenous and *ladino* poor, peasants, permanent and temporary farm workers, migrants, students, produce and grain venders, and rural women. Its strength was its connection to the rural realities and collective forms of living, whether by food self-sufficient communities or rural proletarians. As short trainings began to take place across the country, it became clear that organization, education and mobilization went hand-in-hand: local assemblies were spaces for all to learn the conditions of each community, and local leaders were expected to be examples of humility, determination, solidarity and courage. Rather than study lessons from a textbook, CUC trainings were based on learning from the organization's experience, with the goal of improving its practice. Study and *formación* were emphasized as permanent tasks of CUC cadre (CUC n/d), tasks that CUC would, decades later, emphasize within internationalist Via Campesina spaces.

The peasant and revolutionary organizations in Guatemala, including the guerrilla armies, saw the local organizations in San Martín as representing a 'conscious force' with the ability to take on social, political, ethnic and productive aspects of reality (Camposeco Cruz 2011). As such, the actions of CUC were premised on adding to, and building upon, the political consciousness of local leadership. In the case of promoters and participants in the PtP program, the CUC argued that 'what good is learning about sustainable agriculture if you have no land?' (Atz, in Camposeco Cruz 2011). There was room for mutual learning between the program and the peasant organization, as long as there was a recognition that the program's successes were insufficient in solving the historical structures of inequality and injustice in the municipality. As such, in the 14 communities influenced by the PtP sustainable agriculture program, the position of the communities before the state and society became one of agrarian demands as well as an alternative technological proposal for farm management, based on horizontal exchanges (Camposeco Cruz 2011). Already in 1978, Roland Bunch, coordinator of World Neighbors' PtP program, had found that several of its promoters were also local leaders of CUC.

As CUC leaders began to organize across the country, they developed the capacity to adapt to local conditions. Organizing in peasant communities was not the same as organizing migrant farm workers. It was necessary for organizers to be able talk about the origins of land scarcity, racism and the monoculture plantations where workers faced extreme exploitation. It was also necessary to begin literacy efforts among farm workers, so that they could use labor law to protect themselves. The first pamphlets for political education were produced by CUC in the late 1970s, and focused on key labor struggles such as wages and working conditions (CUC n/d). Subsequent educational material focused on the Mayan historical memory and the right to be free from abuses. CUC educational processes also sought to channel the innate leadership qualities in communities and among workers. In San Martín, local CUC leaders saw PtP as a 'vehicle' for achieving greater level of organization among peasants (Interview MA, 21 August 2017). Thus, already in the late 1970s, rural social movements were considering what would later be known as 'agroecological' organizing structures as a tool contributing to the 'conscious force' of the peasantry.

During the two mayoral terms of Felipe Alvarez in the 1970s, the system of peasant mutual aid called *chuchubal* was incorporated into local policy, vastly increasing the replicability of the first PtP experiences, which became part of the "massification of the construction of soil and water conservation works, schools, roads, and highways" (Camposeco Cruz 2011: 15; translation by authors) carried out under the indigenous-led local government after the earthquake. This may indeed be the first documented scaling-up of agroecology using horizontal methods. It also represents the hightide mark of the PtP method in Guatemala, contributing to the physical recuperation of the lands of San Martín Jilotepeque, but also as a major contribution by Kaqchikel peasants to the cultural and educational content of the Guatemalan revolutionary platform. The philosophical basis of the San Martín PtP experience was the Mayan concept of seeing and feeling the land as a sacred being that must be taken care of, so that it can take care of living plants and these, in turn, can take care of people and animals. Methodologically, PtP in Guatemala was based on 'inductive horizontalism' in the teaching–learning process, moving from the simple to the complex, the small to the large, and the empirical to the systemic (Camposeco Cruz 2011). Solidarity among all the actors within the PtP structure was fundamental to creating a critical mass of promoters and participants that would carry the method on to other municipalities within Chialtenango, and eventually to other countries, despite the monumental state violence they faced.

Felipe Alvarez was kidnapped and disappeared, along with his 17 year-old son, by the military government in 1980. The following year, 1981, three more of his sons were disappeared. Although nearly half the PtP promoters and participants were not associated with the revolutionary struggle, the military repression against them was nonselective, and by 1982 the only promoters who remained in San Martín were those who had been killed there (García 2012). Many PtP promoters fled to parts of Guatemala where the repression was less intense, and those associated with the revolutionary struggle entered the dense mountainous forest controlled by guerrilla organizations. Others fled to neighboring countries. Those who arrived in Mexico were the foundational trainers for long-standing PtP processes in Tlaxcala, Puebla, Oaxaca and Chiapas. In Honduras, the Guatemalan trainers became the subject of the famous farmer-first 'bible', *Two Ears of Corn* (Bunch 1982). However, it would be the subsequent arrival of PtP trainers to Nicaragua, in 1987, that would become highly significant to today's discussion of taking agroecology to scale.

Meanwhile, the CUC became a primary target of the Guatemalan military government's repression. In January 1980, during a peasant occupation of the Spanish embassy to protest oil exploration in Quiché, Guatemalan security forces invaded the embassy using flamethrowers and white phosphorus grenades. Of the 38 people who burned to death, 21 were CUC members. Within two years, the 'scorched earth' policy of genocide had decimated the CUC's national leadership, and the organization ceased to exist publicly. The use of paramilitary 'self-defense' patrols, the forced migration to 'model villages' and the systematic disappearances of community leaders, were pillars of the counter-insurgency strategy of war against popular organizations (CUC n/d). Only in March 1985, in Mexico, did the CUC begin to reorganize, its primary national and regional leaders having been killed or disappeared in the repression.

Nicaragua: cauldron of peasant action

According to Felipe Tomas, Kaqchekel peasant cofounder of PtP in San Martin, the methodology arrived to Nicaragua brought not directly by its Mayan founders, but rather by their disciples, the Tlaxcala indigenous farmers of Central Mexico (interview, October 2016). The methodology 'fell into the lap' of peasant organizing efforts in the heat of the Sandinista Revolution, the ideology of which mixed Marxist political economy with Christian liberation theology and the radical democratic character of Freirian popular education. The National Union of Farmers and Ranchers (*Unión Nacional de Agricultores y Ganaderos*—UNAG) had been created in 1981 by the Sandinista Front in order to represent the interests of small, medium and some large farmers. This interclass rural alliance of landholders played a key role in communicating the goals of the Revolution and the Sandinista Agrarian Reform in the countryside, as well as bringing medium farmers' demands to the table of the Sandinista leadership. Small farmers and their interests were also present, although they had more difficulty being represented in the political structure of the UNAG. The *Programa Campesino-a-Campesino* (PCAC), which focused entirely on small farmers, was seen as an interesting program, but not a strategic economic or political effort by the UNAG leadership, and was funded by the Ford Foundation (Holt-Giménez 2006).

Within 3 years, the Soviet Union was disintegrating and in 1990, the Sandinista Revolution peacefully handed over power to a neoliberal government after losing an election in which the opposition, openly supported by the Bush Administration, promised an end to the CIA-sponsored Contra war if elected (Chomsky 1990). A flood of international money to fund democracy in Nicaragua soon came to dominate the national economy, eclipsing the domestic product (Núñez Soto 1995). The international NGO SwissAid assumed the financing of the program and has continued to sponsor it up to the present. After 1990, PCAC ballooned in size and geographic reach, eventually coming to include up to 6000 promotors and 20,000 indirect participants. Thus, in Nicaragua, the PtP method made its second appearance as a mass-action, peasant-led, movement form of agroecology, this time at a national scale.

It should be noted, however, that the two historic cases of PtP discussed above can be contrasted in terms of their political context. While both took place in Mesoamerican popular struggles in which a radicalized tradition of Christian social action merged with Freirian pedagogical concepts, the rest of the panorama was highly distinct in each case. In Guatemala, PtP emerged first as an alternative to, then as a component of, the revolutionary proposal of land reform and the reversal of historic injustices against the indigenous population. It was clearly a process of *concientización* ("consciousness-raising") that led peasant farmers to pose and solve problems, in the midst of a polarizing civil conflict.

In the Nicaraguan case, by contrast, the mushrooming of PCAC took place in the vacuum of policy toward the countryside following the end of the Sandinista Revolution, in a context of privatization, commoditization and co-optation of what had been revolutionary, empowering processes (Núñez Soto 1995). The relatively uninterested stance of the UNAG, and the more hands-on role of SwissAid, in the PCAC of the neoliberal years may have resulted in a form of PtP more program-like, and markedly less movement-like in Nicaragua (Holt-Gimenez 2006). This is evidenced by the simultaneous participation by PCAC promotors in a wide variety of NGO-sponsored trainings during the 1990s and 2000s, many funded by the Interamerican Institute for Cooperation in Agriculture, which focused on teaching peasant farmers to become service providers through ecotourism and other commercial activities, such as trainings and sales of worm compost. Tellingly, these trainings paid farmers per-diems for their attendance, effectively erasing the foundational free service and mutual aid ethics of the San Martin PtP promotors.

Although PCAC still exists in Nicaragua, its scale and prominence have significantly declined. There are very few continuing exchanges organized by UNAG to provide opportunities for one peasant farmer to teach others. However, the farmers who did become promotors through the PCAC almost universally continue to produce using agroecological methods, which is not only evidence of the soundness of the production techniques, but also of the level of knowledge and commitment reached by farmers who do not just practice agroecology but also teach it. Despite facing challenges such as novel diseases and pests, price fluctuations and an aging workforce, agroecological farmers 'left to themselves' have continued to diversify and integrate the components of their agroecosystems, creating 'beautiful farms' (Ploeg 2008). In several cases, these farmer-promotors have been able to connect with other institutions or organizations (such as the Nicaraguan farm extension institute, the National Agrarian University, or Via Campesina member organizations) to continue sharing their knowledge with would-be agroecological farmers. In these cases, the institutions and processes looking to use original PCAC promotors have found ways to 'de-commodify' the exchanges, either by inviting the promotor and family to other activities as a trade for informally receiving groups, or by promising to buy fruits and vegetables from farmers who host exchanges.

Method in movement: peasant-to-peasant in Cuba

In 1996, Leonardo Chirino, a leading cadre of the Cuban small farmer organization, ANAP (*Asociación Nacional de Agricultores Pequeños*), was traveling through Nicaragua to Honduras but was temporarily denied a visa by the Honduran authorities (Multimedia Peasant School 2014). Looking to keep him busy for the unscheduled days he spent waiting in Nicaragua, UNAG personnel introduced Chirino to the PCAC experiences in Boaco. Upon returning to Cuba, his report on the vast potential for the method in Cuba led the ANAP to seek funding for a pilot PtP program in the central province of Villa Clara. After showing exponential growth during 2 years, the national ANAP leadership decided to throw the political will of the organization behind the PtP method, and in 2001, ANAP president Orlando Lugo announced the creation of a national movement within ANAP, called the *Movimiento Agroecológico Campesino-a-Campesino* (Rosset et al. 2011).

The historically unique success of agroecological production in Cuba has been the source of fruitful scholarship (for example Rosset and Benjamin 1994; Wright 2005; Machín et al. 2010). The main driving factor for Cuba's transition to agroecological farming was undoubtedly the "Special Period in Time of Peace" which began in the early 1990s as agricultural chemicals formerly imported from the Socialist Bloc suddenly dropped out of sight, and all sectors, including the large state farm sector, were forced to make do with less inputs. However, compared to the state and collective sectors, the peasant sector in Cuba has displayed unique characteristics in rebounding back and exceeding pre-Special Period production levels, through the application of ecological principles (Reardon and Aleman 2010; McCune et al. 2011). The *Movimiento Agroecológico Campesino-a-Campesino* has been the fundamental tool for transforming isolated experiences by diligent ecological farmers into widespread, massive agroecological learning processes with strong social momentum (Machín et al. 2010), making

Cuba's food system increasingly resilient and self-sufficient despite ongoing challenges (Chan and Freyre Roach 2010).

The transformation of PtP, from a pilot program in Villa Clara into a national movement within the Cuban Revolution, was the key moment in moving beyond the structural limitations of the PCAC as inherited from Nicaragua's UNAG (see Table 2). For example, rather than a program that maintains a specific staff within an organization, the 'movement form' of PtP required that every ANAP cadre in the country assume responsibility for the development of the *Movimiento Agroecológico Campesino-a-Campesino*. This movement became an 'organic task' of the ANAP, and thus independent from the cycles of external NGO funding (Machín et al. 2010; Rosset et al. 2011). In 2009, Via's Working Group on Sustainable Peasant Agriculture (now reorganized into the Collective on Agroecology, Seeds and Biodiversity) sent a team to 'systematize' (document, analyze and synthesize) the experience of the *Movimiento Agroecológico Campesino-a-Campesino* in Cuba, in order to pull out the best lessons for other Via members wishing to take agroecological farming to scale (Rosset et al. 2011). Another such team was created to systematize the Zero Budget Natural Farming movement in India (Khadse et al. 2017). These processes of systematization, as well as international courses on the PtP method for Via educators at the ANAP national school in 2013 and 2015, have essentially released the PtP method, 'from Cuba to the world'. Over the next decade, the cultural translation and appropriation of PtP principles in peasant movements across the globe may show just how scalable the method proves to be, beyond the specific historical circumstances discussed here and in other PtP literature.

Territorial learning across the continent: Latin American Agroecological Institutes

As mentioned above, a significant part of the Guatemalan peasantry became a diaspora during the 1980s and 1990s. In the case of PtP promotors, these people became carriers of a fundamentally new form of what would come to be called agroecological education, based on the radical premise that the peasants are the best teachers of peasants. The phenomenal voyage of PtP (summarized in Table 3), and its transformation of farming systems in Mesoamerica and the Caribbean, have been well-documented (Holt-Giménez 2006; Machín et al. 2010) and shown to have created massive territorial impacts as ever-larger numbers of peasants have implemented agroecological techniques.

On the other hand, the story of how the Mayan strand of popular, political education as practiced by CUC in the 1970s became interwoven with other movements' educational processes and eventually transformed into the continental articulation of Via schools, is much less

Table 2 Characteristics of PtP in several countries. (*Sources* Camposeco Cruz, 2011; Holt-Giménez, 2006; Machín et al. 2010)

Organization–Country	Global Neighbors—Guatemala	UNAG—Nicaragua	ANAP—Cuba
Time period	1972–1979	Since 1987, but height of 'movement form' from 1989 to 1999	Since 1999, but taken 'movement form' since 2001
Type of organization	US-based NGO connected with Catholic base communities in Guatemala	National organization of small, medium and large producers	National organization of small producers, mostly beneficiaries of agrarian reform processes
Political-historical context	Civil war, polarization of society	Revolutionary and neoliberal eras	Special period, food crisis
Type of PtP process	Small, local program funded by international NGOs	Large, national program funded by international NGOs	Large, national movement (independent of NGO funding) within ANAP and Cuban Revolution
Promotors	Local experimenters who covered large areas, traveling to train farmers through practice	Local experimenters who mostly receive other farmers on the parcel of promotor	Local replicators (experimentation is less emphasized) who receive other farmers on the parcel of promotor
Participation and motivation of promotors	Voluntary, based on Mayan mutual aid; Catholic concept of service to society	Voluntary in first decade, recently paid; combination of moral and material motivation	Strictly voluntary; based on patriotic and revolutionary concept of the peasant's role in society, as well as the Cuban system of social recognition
Facilitators	Almost not mentioned, but could include the trainers at Rafael Landivar University as well as Global Neighbors staff	Clearly defined role for UNAG cadres	Clearly defined role for ANAP cadre or locally recruited cadre of each cooperative
Schools	Mobile schools of soil and water conservation, in each community	Not part of the methodology	Deep connection with local primary schools; coordinators and cooperative leaders are trained at the Niceto López National Training Center in Artemisa

Table 3 Comparing CUC and *campesino-a-campesino* in Guatemala of the 1970s

Name	CUC	CaC (PtP)
What	Social-political organization	Methodological process of exchange
Who	Men, women, youth, farmworkers and peasants	Strictly landholding peasant farmers
Where	Across Guatemala	Municipality of San Martín Jilotepeque
Why	National transformation and end to oppression	Mayan *chuchubal*, Catholic ethic of serving community and sharing knowledge
Concept of cadre	Humble, transparent, courageous community leader	Good farmer willing to share and lead by example
Concept of training	Studying the historical moment and organizational methods, collectively reflecting upon lived experience	Learning-by-doing agroecological practices on the parcel
Concept of school	Space for temporarily removing cadres from their reality to reflect upon it and become more complete revolutionaries	Parcel with several agroecological practices and a farmer-trainer to learn from

documented and poorly understood, given its importance in recent Latin American history. This section will explore the roots and principles of the Latin American Agroecological Institutes, internationalist agroecological training schools of the *Coordinadora Latinoamericana de Organizaciones del Campo* (Latin American Coordination of Rural Organizations, the continental expression of Via, known as CLOC), which bring together peasant youth from Puerto Rico to the Patagonia. We illustrate the necessity of a theoretical framework that integrates territorial and generational challenges into the praxis and pedagogy of agrarian movements. Unlike PtP, the CLOC's schools have no single common origin; instead, their roots are in a shared 'historical memory' of indigenous, Black, worker, peasant and women's resistance.

It is useful to recognize two eras in radical pedagogy, with the publication of Paulo Freire's *Pedagogy of the Oppressed* in 1970 marking the turning point between the two. Analyzing the origin of the political education of Via—or indeed PtP—would be impossible without recognizing the long tradition of critical theory and pedagogy in Latin America, including Simón Rodríguez, the teacher and mentor of Venezuelan and continental liberator Simón Bolívar, as well as José Martí, whose pedagogical thinking inspired generations of revolutionaries. Brazilian educator Paulo Freire himself often referred to Ché Guevara's writings on the Cuban revolutionary struggle as a case of dialectical co-learning between, on one hand, radical social transformers who insisted on the need for armed struggle, and, on the other hand, peasant communities that insisted on the primacy of land amongst the demands of the revolutionary war. From the Worker's University founded by Julio Antonio Mella in the Cuba of the 1920s, to José Vasconsuelos's experiments with rural education in post-revolutionary Mexico, debates about the relationship between education and structural change were present in virtually all Latin American struggles for independence, sovereignty and decolonization.

We find three critical dialectics in this earlier tradition of radical education. First, the tension in class society between manual and mental labor, by which the former is privileged and the latter undervalued. Second, the sequential relationship between action and reflection, by which learners bring their prior knowledge of practical activities to spaces of abstraction, only to return to material reality with refreshed ideas. Third and finally, the dialectical relationship between the transformation of the social world and the transformation of the individual who, in the process of changing the world, changes him- or herself. Manual-mental, action-reflection, individual-world: these are the interconnected vertices of transformation in classic Latin American critical pedagogy (Pérez 2009). Developing this tradition, Freire (1970) interrogated a fourth dialectic: the relationship between teacher and student, *educador* and *educando*, arguing that the power relationships within educational spaces is part of the system of oppression that denies the oppressed their right to both action and word. While conventional education objectifies the learner, equates him or her to a container that needs to be filled, liberatory education poses questions that only collective reflection and action can solve, in a historical process known as praxis.

Paulo Freire's exile from Brazil and the subsequent proliferation of popular education in Latin America and Africa produced hundreds of formal and informal educational settings that sought to implement the pedagogy of the oppressed, as well as integrate feminist theory into educational praxis. In countries such as Mexico, Chile, Argentina, Nicaragua, Cuba, Guinea-Bissau, Angola and many others, popular education became integrated with the ideas of the progressive elements within the Catholic Church, developing significant territorial processes of emancipatory demands and consciousness-raising by poor people. Indeed, in Nicaragua, the popular education of the 1980 literacy campaign symbolized the ethos of the Sandinista Revolution—popular, socialist and Christian (Moro 2004). Rather than starting from abstract concepts of European critical thought, these territorial processes of popular education have emphasized the construction of a critical attitude toward daily life, taking learners' living experience as a subject matter to be studied. The 'option for the poor' that embodied Liberation Theology was a departure from the determinism of Eurocentric Marxism, and inspired creative, emancipatory action and ideas, including action research (Fals Borda 1978). The process of recognizing, and opposing, the structures of oppression that are reproduced in daily life, is considered a historical and pedagogical task by Via, especially through its Global Campaign to End Violence against Women.

The social movements of Latin America embody this tradition of struggle based upon daily experience rather than dogma, especially in the period since *Pedagogy of the Oppressed* was published in 1970. Particularly in the countryside, movements' autonomous educational processes are centered in the spheres of sociocultural production and reproduction. As such, the educational systems they create are embedded within their food systems and seasonal labor patterns. This is a break from the conventional approach to education, which separates knowledge and context (Giroux 1981). On the encampments of the Landless Workers' Movement of Brazil, autonomous movement schools are operated using Freirian conceptions and methods, arranged to complement collective work in fields (Caldart 2002). Similarly, CUC current educational processes, including its National School for Cadres (*Escuela Nacional de Cuadros*) are set into existing seasonal and daily patterns of farming. Similar patterns exist in the political-agroecological training schools of Via and CLOC across the continent (McCune et al. 2014; Muñoz et al. 2014).

In 2005, Venezuelan President Hugo Chavez signed a landmark agreement with the Brazilian Landless Workers' Movement to build an international school of agroecology in Barinas, the rural home state of Chavez. Over the next several years, a group of 110 young people, children of peasant farmers from across Latin America, took part in the educational and productive processes of the Paulo Freire Latin American Agroecological Institute (*Instituto Agroecológico Latinoamericano*, known as IALA Paulo Freire) (McCune et al. 2014). A continental process of exchange among organizations led to the creation of a network of movement agroecology schools, in which agroecology cannot be separated from the political conflict between peasant agriculture and corporate landgrabbing. As of January 2018, Latin American Agroecological Institutes exist in Argentina, Brazil, Chile, Colombia, Ecuador, Nicaragua, Paraguay, and Venezuela, with several others in process.

In these schools, both younger and older educational subjects are trained not only in farming, but also in political-organizational aspects of the collective struggle for land reform, food sovereignty and agroecology. This makes for a rich learning environment, as learner-educators and educator-learners embark on dialogue-based processes of action and reflection, using popular education as a guide for developing critical consciousness. In the Latin American Agroecological Institutes, social movement members are selected by their organizations to study for long periods of time (3–5 years) so they can return and play a leading role in building agroecology as a historical replacement to the capitalist agro-food system. The philosophical principles of these movement schools are shown in Table 4.

Movement agroecology schools are motivated by the recognition that structural change is necessary for agroecology to solve hunger, water scarcities, and injustices due to ongoing climate change. Rather than only teach the productive aspects of agroecology, these schools develop critical thinkers (Muñoz et al. 2014). 'Teach me how to think, not what to think' could be an abbreviated version of the critical approach to agroecological education. An integral education means that people understand the ethical, civic, political, economic, ecological and social basis for agroecology. This helps them become collective actors, in the form of rural social movements, capable of moving forward the agenda of structural change so that food sovereignty can be a reality (Moore 2017b).

Recovering historical memory is another strategic objective of Via's peasant movement schools. Due mass media bombardments, superficial stereotypes and consumerism, it is common for young people to know little about their roots. In movement schools, educators retrace the histories of popular struggles, such as indigenous resistance to displacement, Black resistance to slavery, peasant resistance to modernization schemes, and women's resistance to objectification and violence, to put social movements' current efforts within a historical frame. Historical memory is necessary for constructing collective identity, and, in many cases, developing the self-esteem of students. The pedagogical principles of the Latin American Agroecological Institutes are shown in Table 5.

Movement agroecology schools are about creating social situations in which learners can transform their way of thinking and doing, so that together they can create the microcosm of the society they wish to live in. This means that each person is responsible for building the school community, rather than leaving everything to a group of administrators. Decisions are made collectively and learner-educators become accustomed to taking responsibility for their actions. Human qualities such as humility, honesty, integrity and solidarity are considered as important to the learning process as are composts, intercropping, and seed saving (Moore 2017b).

Movements employ the concept of 'social situations of development', developed by Soviet psychologist Lev Vygotsky (1978). Learner-educators use their cultural medium to internalize knowledge that first exists on an interpersonal

Table 4 Philosophical principles of the Latin American Agroecological Institutes. (*Source* Muñoz et al. 2014)

Education through and for social transformation
The development of women and men with new values as well as new emotional linkages to others, resulting in actions directed at social transformation, opting always for the people and rejecting lifestyles promoted by neoliberalism. Included here are the most elevated of human values needed for subjects taking on their own agroecological education, including solidarity, humility, equality, justice, honesty, internationalism, and respect for nature, among others
Education though and for diversity
Neoliberalism promotes a sole culture in which all people are expected to reproduce the anti-values of consumerism, domination, and selfishness. Agroecological education, on the other hand, recognizes and promotes the indigenous, African, feminist, anti-colonial and anti-imperialist struggles that have accompanied our people for over 500 years. Agroecologists stand opposed to that dominant culture, defending instead the enormous amounts of cultural diversity found in popular human systems as well as the biodiversity used by Mother Earth to organize our planet
Education through and for work and cooperation
Work is understood as a means by which women and men dignify their existence. Work is considered a form of liberating action instead of a commodified need of working people. Studying is directly linked to productive efforts through work and volunteering, with both these actions considered a means by which the world can be better understood. Cooperation is used so that new citizens educate themselves collectively, developing the capacity to collaborate through a democratic dialogue. Cooperation becomes an ethical necessity in both work and study and is present in processes between students themselves, between students and popular educators, and between, students, popular educators, and communities
Education through and for rebellion
Citing Paulo Freire, "we struggle for an education that teaches us to think—not one that teaches us to obey". Agroecological education in this context openly questions and confronts social injustice, while at the same time directing students' efforts into collective processes of social transformation that have at their heart humanity's pending humanization. Rebellion is promoted so that a better world becomes reality

Table 5 Pedagogical principles of the Latin American Agroecological Institutes. (*Source* Muñoz et al. 2014)

Practice-theory-practice
For popular education to exist, acts of praxis are constantly taking place based on a reciprocal relationship of dialogue between action, reflection, and matured action. True education takes place when society is being transformed
Teaching–learning
A dialectical and horizontal relationship exists between educators and learners, with both teaching and learning in a constant dialogue free of hierarchy. Educating and learning come together in one single act of education, "forming" collectives of people committed to their social responsibilities. Every member of the educating community commits themselves to each other's learning, taking full advantage of the time and space available to harvest the greatest amount of education possible.
Dialogue among ways of knowing
Convinced that only through a diversity of visions, perspectives, and proposals do people come to truly understand the world around them, a real communication is built between participants that allows for the free flow of knowledge, ideas, feelings and awareness, recognizing the conceptual legitimacy of all those who struggle for a better world
Action-based, participatory, and contextualized research
Investigations that take place are directly related to the real needs of students, their families and communities. Never are people, peasants in this case, considered the objects of academic research. Rural people and their organizations, with special attention paid to the youth, are the protagonist subjects of all inquiry developed to achieve both education and liberation. In addition, all research has an overriding strategic objective—contributing to food sovereignty.

plane before being repatterned and appropriated by the structures of the individual mind (1978). Table 6 shows how distinct pedagogical activities promote learning at an individual, interpersonal and social historical level within social movement political-agroecological training schools (McCune et al. 2017). The socio-educational dimension of agroecological scaling produces 'situations of territorial development': contexts produced by the social movement with the purpose of stimulating critical processes of thought and action for peasant, peasant-indigenous and peasant-worker territories. In such a way, learning clearly transcends the individual mind and occurs in groups, institutions, and, most importantly, places (Basso 1996).

As Table 6 shows, emphasis in Via's agroecological schools is placed on opportunities for creative thinking, engagement with ideas outside of traditional calculations and essays, and metalearning by which subjects are conscious and critical of their own development process. These activities have distinct results at different levels—while individuals may become more sensitized and capable of dialogue, movements become more integrated and agile. But some of the results are similar across levels: communities

Table 6 Three levels of learning and knowledge mediated by distinct pedagogical activities at the Escuela Obrera Campesina 'Francisco Morazán' in Nicaragua. (Source: McCune et al. 2017)

Activity	Level		
	Individual subjects	Interpersonal: small groups and classroom	Social historical subjectivity
Mística	Sensitizing and motivating learner	Realization of diversity potential, rescue of historical memory	Re-production of traditional culture, re-valuing disrespected forms of knowing
Permanent small groups with names and slogans for dividing chores at school	Learning to work with different types of people	Achieving fuller participation	Generates more diverse 'answers' or representations within unity of collective subject
Gardening: double-digging, seed selection, soil preparation, composting, thinning, transplanting, watering, mulching, combining crops, weeding, rooting, harvesting, tasting	Connecting practice with new information, building on previous knowledge, discovering new interests, questions, and qualities in oneself	Produces bonds among 'co-workers', pride, recognition of previous knowledge	Gaining ability to be food self-reliant, and teach production methods for generating food sovereignty
Readings and analysis of texts	Strengthening reading and synthesis skills	Fomenting debate, stimulating zone of proximal development	Reinforcing historical memory, employing relevant categories of analysis
Physical-emotive games	Motivations, stimulating creativity	Stimulating self-esteem, collective bonds, interpersonal trust	Creating new categories of meaningful fun, bonding
Exchanges with local peasant farmers and communities	Developing protocol and methods for working in communities	Breaks out of the classroom dynamic, improves reflection	Legitimizes course in eyes of the community; broadest contextualization of Via's courses
Theater created by *educandos* in small groups and performed for class and guests	Creative thinking, transforming course itself into 'performance' to observe critically	Gaining fuller participation, trust within group, bringing 'electricity' to group	Gaining a simple, flexible 'tool' for critical reflection in communities
Interpretation of poetry and songs	Use of figurative language, cultural codes	Chance to share and compare taste in music and art	Connection between historical moment and the art it produces
Intergenerational and intercultural dialogue	Gratifying experience, listening skills	More complete collective constructions and senses	Continuity, *relevo*
Oral and written self-, co- and hetero-evaluation and synthesis of course	Critical writing and thinking skills	Distribution of responsibility for collective processes	Historical record

become more sensitized and capable of dialogue, and there is an exchange of terminologies and language between the territory and the school.

One of the major characteristics of movement schools is their organicity, a term meaning that people are connected to one another in bonds of reciprocity, communication, planning and follow-up of tasks that are defined collectively. Learner-educators work in permanent small groups that share productive, academic and managerial duties within the school setting. Additionally, working groups are established by topic to make sure that all the needed actions at the school are carried out. This form of direct democracy is reflected and improved upon over time through the actions of the Political-Pedagogical Coordination, a body made up of leading cadres from the schools' participating social movements. Coursework is transdisciplinary, combining several academic disciplines with long homestays in peasant communities near the schools, allowing learner-educators to learn from the 'chalkboard of reality' as they develop skills related to rural assessments and participatory community work. This principle of itinerant education considers school time to be only one part of learning; community time is just as important for learners to analyze using similar categories as they use to understand their learning process at the school.

Discussion

Given that peasant economies order and re-order the relationship between society and nature in a different way than capitalist economies (Chayanov 1986), the two kinds of economies create qualitatively distinct kinds of territories. Today, capital's way of ordering territories depends on the accelerated expulsion of peasant and indigenous populations and the incorporation of agribusiness resources (water, land, minerals, animal bodies, etc.) into global values chains that fatten the financial sector while desecrating and polluting actually existing places (Ploeg 2008; Martínez-Torres and Rosset 2014). The way that peasant communities order territories is based on more complex metabolic interactions between biological populations, and, in the case of indigenous peasants, has often shown to be sustainable for hundreds and even thousands of years (Altieri and Toledo 2005). Given such intractable differences between models, the territories constructed in the rural world have class orientation, which leads to protracted political struggle over territory (Fernandes 2017). Social movements are increasingly explicit about creating educational processes for the construction, maintenance and defense of peasant, peasant-indigenous and peasant-worker territories (La Via Campesina 2015). In doing so, gender and ecological relations are revalued, as *reproduction* of the territory becomes more important than 'production', the latter so often linked to capitalist expropriation and the devaluation of the territory.

In such an undertaking, social movements seek alliances with as many sectors of society as possible, in order to create a 'historical bloc' capable of definitively repelling capital's world-ecology (Moore 2017a). Territorial education is part of a Gramsci (2011) 'war of positions' that rural social movements wage against the capital system. As such, peasant pedagogy defends not only material territories but also immaterial territories (Fernandes 2017), including institutions, communications media, and mobilizing concepts like 'local food systems', 'agroecology', 'feminism' and 'food sovereignty'. This education recognizes the diversity of popular forms of knowledge and uses the *diálogo de saberes* (Martínez-Torres and Rosset 2014) to create new senses of struggle. The environmental ethic and the social ethic—artificially delinked in Western thought since Enlightenment (Moore 2017a, b)—are synthesized in a proposal for socio-ecological relations built on collective rights and responsibilities, with popular sovereignty as a condition for food sovereignty. The 'humanization' that Freire (1970) puts in the center of education is no longer only a transformation of the relations among people, but also between people and all other forms of life (Meek and Lloro-Bidart 2017). Agroecological training or *formación*, as understood by Via, adds to the tradition of critical popular education in Latin America by interrogating the dialectic between humanity and living nature.

In this sense, scaling-out agroecology is really a process of territorialization, due to the specificity of agroecological farming. Each agroecosystem is particular, and biological populations often behave entirely differently due to the unique combination of biotic and abiotic conditions. As such, learning processes do not only create movements, but rather 'movement-places'. The organizational coherence, or 'organicity' (Rosset 2015) of the movement—which itself is also a unique arrangement of mental and manual labor—combines with specific ecological structures to produce and reproduce peasant territories that incorporate seeds, water, soil and local knowledge. This meeting of movement and place produces the peasant way of life.

Conventional schooling either excludes peasant youth or teaches them to be anything but peasants. Rural social movements in Latin America have built and continue to build solid alternatives for teaching their cadre—including informal, territorial processes as well as autonomous schools of critical thinking. In doing so, movements are building peasant, peasant-indigenous and peasant-worker territories. The schools—each its own unique historical-cultural process—support the reproduction of territory by providing cadres uniquely prepared to facilitate the relationships among the rest of the components of the movement-territory, such as peasant farmers, public institutions, neighborhood organizations, labor unions, farmers' markets, agroecological techniques and land (McCune et al. 2016). Solidarity is an emerging property of the school-territory system, as economic and cultural co-penetration, relationships of mutual trust, seed sharing and ideological commitments take place in concrete places and times, through actions by real people and ecologies. Food territories are relational, historical and interactive (Schiavoni 2016).

Using agroecology to rebuild the world's food systems requires the joining of cultural, political, economic, social, historical and ecological processes. For agroecology to replace the agribusiness model, a major power shift is necessary—knowledge is not enough to stop what Brazilian geographer Bernardo Mançano Fernandes (2017) calls "capital's most intense resource-grab in history" that began a decade ago with the crisis in food prices. Class struggle is a necessary historical pillar of food sovereignty. However, it is not class struggle per se but rather the 'class project' of a historical bloc or constellation of social sectors linked to specific territories that provides the alternative to corporate, placeless 'food empires' and State acquiescence (Ploeg 2008). The class project of transnational capital and the powers in its orbit is one of expanded capital reproduction at the cost of vastly accelerated resource extraction and commodity value chains leading to the financial sphere. A revolutionary class project, agroecology with territorial defense and redistributive resource devolution, creates movement-territories and new social situations, such as schools, health clinics, community centers and cooperatives, that provide opportunities for organizing food and life differently even in the midst of unavoidable climate change and continued social crisis.

The contributions of agroecological thought and action to class struggle are only beginning to be applied globally, through the coordinated actions of movements such as Via. Gramscian social theory provides a lens for analyzing the weapons of the weak (Scott 1985) in their struggle against globalized capital. Among the multiple shapes that agroecology takes, the 'movement form' of agroecology—as PtP and as movement schools—concretely supports the construction of peasant, peasant-indigenous and peasant-worker territories. As pedagogical processes develop, new territorial dynamics take shape, revealing that self-organized peasant education contributes to sustaining popular territories against the aggression and violence of globalized capital. Time will reveal whether these movement-places are able to develop into counter-hegemonic historical blocs, able to guarantee food sovereignty and resist reproducing capital dynamics. Much depends upon the alliances built between urban and rural popular movements, as well as the dialogues built among the types of knowledges that support them both (Declaration of Auskal Herria 2017).

Learning theory needs to broaden the concept of subject, to include collective and place-based actors. Movements learn and territories learn, in ways that are only beginning to reveal themselves. The specificity of interactions between places and their people, as well as the material flows of social reproduction, appear to be important mediators of the kind of learning needed for territories to flourish. Using the concept of learned territories permits us to conceptualize learning and territorial construction as two facets of one historical process.

Acknowledgements This study would have been impossible without the work, sacrifice and teachings of hundreds of thousands of small farmers who have participated in the peasant-to-peasant movement, as well as the peasant organizations that make up La Via Campesina, including CUC, ATC and ANAP. The authors would like to acknowledge the support and guidance of Faustino Torrez, Peter Rosset, Julia Margarita Trujillo, Rilma Román, Adilén Roque, María Amalia Rodríguez, Leonardo Chirino, Felipe Tomás, María del Rosario Valenzuela Sotomayor, Eric Holt Giménez, Daniel Pascual, Rafael González, Gregorio Chay, Juan Reardon and Margarita Fernández, as well as the invaluable suggestions of Colin Anderson and three anonymous reviewers for improving this manuscript.

References

Albizures, M. A. 2008. El CUC en la historia. Opinión., ALAI Online. Retrieved from 16 Apr 2008.

Althoff, A. 2014. *Divided by Faith and Ethnicity: Religious Pluralism and the Problem of Race in Guatemala*. Berlin/Boston: Walter de Gruyter.

Altieri, M. A. 1989. Agroecology: A New Research and Development Paradigm for World Agriculture. *Agriculture, Ecosystems and Environment* 27: 37–46.

Altieri, M. A., and V. Toledo. 2005. Natural Resource Management among Small-scale Farmers in Semi-arid Lands: Building on Traditional Knowledge and Agroecology. *Annals of Arid Zone* 44 (3): 365–385.

Barbosa, L. 2013. Pedagogías Alternativas: el Movimiento de los Sin Tierra y el Movimiento Zapatista. In *Reconfiguración de los Educativo en América Latina*, eds. M. Gómez Sollano, and M. Corenstein. Mexico City: UNAM.

Barbosa, L. P. 2015. *Educación, Resistencia y Movimientos Sociales: la práxis educativo-política de los Sin Tierra y de los Zapatistas*. Mexico City: LIBRUNAM.

Barbosa, L. P. 2016. *Educação do Campo* [Education for and by the countryside]as a political project in the context of the struggle for land in Brazil. *Journal of Peasant Studies* 44 (1): 118–143.

Basso, K. 1996. *Wisdom sits in places: Landscape and language among the Western Apache*. Albuquerque: University of New Mexico Press.

Bunch, R. 1982. *Two Ears of Corn*. Oklahoma City: World Neighbors.

Caldart, R. S. 2002. *Pedagogia do Movimento Sem Terra*. Sao Paulo: Expresso Popular.

Camposeco Cruz, M. 2011. La propuesta de desarrollo sostenible denominada de campesino a campesino: Una experiencia en San Martín Jilotepeque: 1972–1982 (Guatemala: Master Thesis, Universidad de San Carlos de Guatemala, Faculty of Political Sciences, May 2011).

Carroll, C. R., J. H. Vandermeer, and P. M. Rosset. 1990. *Agroecology*. New York: McGraw-Hill.

Chan, M. L., and E. F. Freyre Roach. 2010. *Unfinished Puzzle. Cuban Agriculture: The Challenges, Lessons and Opportunities*. Oakland: Food First Books.

Chayanov, A. V. 1986. *The theory of peasant economy*. Madison: University of Wisconsin Press.

Chomsky, N. 1990. *The Decline of the Democratic Ideal*. April 1990 Z Magazine.

Comisión para el Esclarecimiento Histórico 1999. «Guatemala: Memoria del silencio» (edición en linea - AAAS). Guatemala: CEH. Archivado desde el original el 5 de mayo de 2013.

Copeland, N. 2018. Meeting peasants where they are: Cultivating agroecological alternatives in neoliberal Guatemala. *The Journal of Peasant Studies*. https://doi.org/10.1080/03066150.2017.1410142.

CUC. (No Date). *Lucha, Resistencia e Historia*. Guatemala: Comité de Unidad Campesina.

De Schutter, O. 2010. Report submitted by the Special Rapporteur on the right to food, Olivier De Schutter, United Nations Human Rights Council, 16th session, agenda item 3.

Declaration of Auskal Herria. (2017). Final Declaration of the 7th International Conference of La Vía Campesina. 22/7/2017. Derio, Auskal Herria.

Declaration of Nyeleni. (2015). *Declaration of the international forum on agroecology*. Nyéléni, Mali, February 27.

Dewey, J. 1915. *The school and society*. Chicago: University of Chicago Press.

Fals Borda, O. 1978. Por la praxis: el problema de cómo investigar la realidad para transformarla, en Simposio Mundial de Cartagena, Crítica y política en ciencias sociales, Bogotá, Punta de Lanza-Universidad de Los Andes, vol. I, 209–249. 1978.

Ferguson, B. G., and H. Morales. 2010. Latin American agroecologists build a powerful scientific and social movement. *Journal of Sustainable Agriculture* 34: 339–341.

Fernandes, B. M. 2017. Territorios y Soberanía Alimentaria. *Revista Latinoamericana de Estudios Rurales* 1 (4): 22–38.

Francis, C., G. Lieblein, S. Gliessman, T. A. Breland, N. Creamer, L. Harwood Salomonsson, J. Helenius, D. Rickerl, R. Salvador, and M. Wiedenhoeft. 2003. Agroecology: The ecology of food systems. *Journal of Sustainable Agriculture* 22 (3): 99–118.

Freire, P. 1970. *Pedagogy of the Oppressed*. New York: Herder and Herder.

García, G. 2012. Las guerrillas y las mayas: una aproximación a las formas de interacción sociopolítica entre las insurgencias y los kaqchikeles de San Martín Jilotepeque (1976–1985). In *(Coord.) Guatemala: la infinita historia de las resistencias*, ed. M. Castañeda, Guatemala: Magna Terra.

Giraldo, O. F., and P. M. Rosset. 2016. La agroecología en una encrucijada: Entre la institucionalidad y los movimientos sociales. *Guaju* 2 (1): 14–37.

Giroux, M. A. 1981. *Ideology, culture and the process of schooling*. Philadelphia: Temple University Press.

Gliessman, S. 2007. *Agroecology: Ecological processes in sustainable agriculture*. Boca Ratón: CRC Press.

Gliessman, S., and M. A. Altieri. 1982. Polyculture cropping has advantages. *California Agriculture* 4: 15–16.

Gramsci, A. 2011. *The prison notebooks*. New York: Columbia University Press.

Holt-Giménez, E. 2006. *Campesino-a-Campesino: Voices from Latin America's Farmer to Farmer Movement for Sustainable Agriculture*. Oakland: Food First Books.

Holt-Giménez, E. 2010. Linking farmers' movements for advocacy and practice. *The Journal of Peasant Studies* 37 (1): 203–236.

Holt-Gimenez, E., and A. Shattuck. 2011. Food crises, food regimes and food movements: Rumblings of reform or tides of transformation? *The Journal of Peasant Studies* 38 (1): 109–144. https://doi.org/10.1080/03066150.2010.538578.

Khadse, A., P. Rosset, H. Morales, and B. Ferguson. 2017. Taking agroecology to scale: The zero budget natural farming peasant movement in Karnataka, India. *The Journal of Peasant Studies*. https://doi.org/10.1080/03066150.2016.1276450.

Klein, N. 2007. *The shock doctrine*. New York: Metropolitan Books.

La Via Campesina. (2015). *Peasant agroecology for food sovereignty and Mother Earth: Experiences of La Via Campesina. Notebook No. 7*. Harare: La Via Campesina International.

Machín, B., A. M. Roque Jaime, D. Ávila, Lozano, and P. Rosset. 2010. *Revolución agroecológica: el movimiento campesino a campesino de la ANAP en Cuba*. La Habana: Asociación Nacional de Agricultores Pequeños and La Vía Campesina.

Martínez-Torres, M. A., and P. Rosset. 2014. Diálogo de saberes in La Via Campesina: Agroecology and food sovereignty. *The Journal of Peasant Studies* 41 (6): 979–997.

McCune, N., Y. Ruíz González, E. Águila Alcántara, O. Fernández Martínez, C. O. Fundora, N. Castillo-Arzola, P. Cairo-Cairo, M. D'Haese, S. DeNeve, and F. Guevara-Hernández. 2011. Global questions, local answers: soil management and sustainable intensification in diverse socioeconomic contexts of Cuba. *Journal of Sustainable Agriculture* 35 (6): 650–670.

McCune, N., J. Reardon, and P. Rosset. 2014. Agroecological formación in rural social movements. *Radical Teacher* 98: 31–37.

McCune, N., P. Rosset, T. Cruz Salazar, A. Saldívar Moreno, and H. Morales. 2016. Mediated territoriality: rural workers and the efforts to scale out agroecology in Nicaragua. *The Journal of Peasant Studies* 44 (2): 354–376.

McCune, N., P. Rosset, T. Cruz Salazar, H. Morales, and A. Saldívar Moreno. 2017. The long road: rural youth, farming and agroecological formación in Central America. *Mind, Culture, and Activity*, 24 (3): 183–198.

Meek, D. 2015. Learning as territoriality: The political ecology of education in the Brazilian landless workers' movement. *The Journal of Peasant Studies* 40 (2): 1–22.

Meek, D., and T. Lloro-Bidart. 2017. Introduction: Synthesizing a Political Ecology of Education. *The Journal of Environmental Education* 48 (4): 213–225.

Meek, D., and R. Tarleu. 2015a. Critical Food Sovereignty Education and the Question of Race. *Journal of Agriculture, Food Systems, and Community Development* 5 (4): 131–135.

Meek, D., and R. Tarleu. 2015b. Critical Food Systems Education (CFSE): Educating for food sovereignty. *Agroecology and Sustainable Food Systems*. https://doi.org/10.1080/21683565.2015.1130764.

Méndez, V. E., C. Bacon, and R. Cohen. 2013. Agroecology as a transdisciplinary, participatory and action-oriented approach. *Agroecology and Sustainable Food Systems* 37 (1): 3–18.

Moore, J. 2017a. The Capitalocene, Part 1: on the nature and origins of our ecological crisis. *The Journal of Peasant Studies* 44 (3): 594–630.

Moore, S. S. 2017b. Organize or die: farm school pedagogy and the political economy of the agroecological transition in rural Haiti. *The Journal of Environmental Education* 48 (4): 248–259.

Moro, W. 2004. Educación popular: Acercamiento a una práctica liberatoria. In *Romero and Hernández (Eds.)*, 35–57 in. Havana: Editorial Caminos: *Concepción y metodología de la educación popular, Toma II*. Formación en Educación Popular Acompañada a Distancia. Pp.

Multimedia Peasant School. (2014). Multimedia Peasant School, an audiovisual tool to scale-up agroecology. Colectivo Koman Ilel. Accessed 7 Nov, 2017, https://viacampesina.org/en/multimedia-peasant-school-an-audiovisual-tool-to-scale-up-agroecolog/.

Muñoz, A., N. McCune and J. Reardon. Agroecological Formación for Food Sovereignty. 2014. *ZNet Communications*, 6 March, 2014. Accessed 13 Feb, 2018, https://zcomm.org/znetarticle/agroecological-formacion-for-food-sovereignty/.

Núñez Soto, O. 1995. *La Economía Popular, Asociativa y Autogestionaria*. Managua: CIPRES.

Ouviña, H. 2017. Gramsci y los movimientos populares como intelectuales colectivos. Gramsci en América Latina. Accessed Feb 3 2017 at: https://gramscilatinoamerica.wordpress.com.

Parmentier, S. 2014. *Scaling-up agroecological approaches: What, why and how?* Discussion Paper, Oxfam Solidariteit.

Pérez, E. 2009. *La promesa de la educación de los oprimidos. En: ¿Qué es la educación popular?*. Havana: Editorial Caminos.

Reardon, J., and R. Aleman. 2010. Agroecology and the Development of Indicators of Food Sovereignty in Cuban Food Systems. *Journal of Sustainable Agriculture* 34 (8): 907–922.

Román, R., and M. Sánchez. 2015. La agroecología: puntal de la soberanía alimentaria. *América Latina en Movimiento* 502: 21–23.

Rosset, P. 2015. Epistemes Rurales y la Formación Agroecológica en La Vía Campesina. *Revista Ciencia y Tecnología Social* 2 (1): 4–13.

Rosset, P., and M. Benjamin. 1994. *The Greening of the Revolution*. Oakland: Food First.

Rosset, P. M., B. M. Sosa, A. M. R. Roque Jaime, and R. A. Ávila, and Lozano. 2011. The Campesino-toCampesino agroecology movement of ANAP in Cuba: Social process methodology in the construction of sustainable peasant agriculture and food sovereignty. *The Journal of Peasant Studies* 38 (1): 161–191.

Schiavoni, C. 2016. Historical, relational, and interactive perspectives in food sovereignty research. *ICAS Review Paper Series* 7: 1–33.

Scott, J. 1985. *Weapons of the weak: Everyday forms of peasant resistance*. Bethany, Yale University.

Sevilla Guzmán, E., and G. Woodgate. 2013. Agroecology: Foundations in agrarian social thought and sociological theory. *Agroecology and Sustainable Food Systems* 37 (1): 32–44.

Stedile, J. P. 2002. Landless Battalions: The Sem Terra Movement of Brazil. *New Left Review* 15: 77–104.

Stronzake. (2013). Movimientos sociales, formación política y agroecología. *América Latina en Movimiento* 487: 27–29.

Valenzuela, M. R. 2008. *¿Por qué las armas? Desde los mayas hasta la insurgencia en Guatemala*. Buenos Aires: Ocean Sur.

van der Ploeg, J. 2008. *The new peasantries: Struggles for autonomy and sustainability in an era of empire and globalization*. Sterling: Earthscan.

Vygotsky, L. 1978. *Mind in Society: The development of higher psychological processes*. Cambridge: Harvard University Press.

Wezel, A., S. Bellon, T. Doré, C. Francis, D. Vallod, and C. David. 2009. Agroecology as a science, a movement and a practice. A review. *Agronomy for sustainable development* 29 (4): 503–515.

Wright, J. 2005. Falta petroleo. *Perspectives on the emergence of a more ecological farming and food system in post-crisis Cuba*. PhD Thesis, Wageningen University.

Nils McCune is a PhD in Ecology and Sustainable Development graduated from El Colegio de la Frontera Sur (ECOSUR), Mexico. He has over 5 years of experience collaborating with La Via Campesina organizations in Mexico and Central America, and is currently a member of the technical team for agroecological training, ATC-Nicaragua and Research Fellow of the University of Michigan School for Environment and Sustainability.

Marlen Sánchez is an Agroecological Engineer graduated from the Paulo Freire Latin American Institute of Agroecology (IALA-PF) in Barinas, Venezuela. She has been a member of the Rural Workers' Association (Asociación de Trabajadores Rurales—ATC) of Nicaragua for 15 years, and is currently the national coordinator of agroecology and a member of the Political-Pedagogical Coordination of La Via Campesina in Central America.

Agriculture and Human Values (2019) 36:611–626
https://doi.org/10.1007/s10460-017-9780-1

SYMPOSIUM/SPECIAL ISSUE

Food sovereignty education across the Americas: multiple origins, converging movements

David Meek[1] · Katharine Bradley[2] · Bruce Ferguson[3] · Lesli Hoey[4] · Helda Morales[3] · Peter Rosset[5] · Rebecca Tarlau[6]

Accepted: 7 March 2017 / Published online: 22 March 2017

Abstract

Social movements are using education to generate critical consciousness regarding the social and environmental unsustainability of the current food system, and advocate for agroecological production. In this article, we explore results from a cross-case analysis of six social movements that are using education as a strategy to advance food sovereignty. We conducted participatory research with diverse rural and urban social movements in the United States, Brazil, Cuba, Bolivia, and Mexico, which are each educating for food sovereignty. We synthesize insights from critical food systems education and the political ecology of education in analyzing these cases. We compare the thematic similarities and difference between these movements' education initiatives in terms of their emergence, initial goals, expansion and institutionalization, relationship to the state, theoretical inspirations, pedagogical approach, educational topics, approach to student research, and outcomes. Among these thematic areas, we find that student-centered research on competing forms of production is an integral way to advance critical consciousness about the food system and the political potential of agroecological alternatives. However, what counts, as success in these programs, is highly case-dependent. For engaged scholars committed to advancing education for food sovereignty, it is essential to reflect upon the lessons learned and challenges faced by these movements.

Keywords Food sovereignty · Critical pedagogy · Education · Critical food systems education · Political ecology of education · Social movements

Abbreviations

ANAP	National Association of Small Farmers (Cuba)
CaC	Campesino a Campesino (Cuba)
CPATI	Colonia Pirai Agroecological Technical Institute (Bolivia)
EBUFFS	East Bay Urban Farmer Field School (USA)
FW	Food Warriors (USA)
LabVida	Laboratorios para la Vida (Mexico)
MACaC	Farmer-to-Farmer Agroecology Movement (Cuba)
MST	Landless Workers Movement (Brazil)

✉ David Meek
ddmeek@ua.edu

1 University of Alabama, 350 Mars Spring Road, Tuscaloosa, AL 35401, USA

2 University of California-Davis, Davis, CA, USA

3 ECOSUR, San Cristóbal de Las Casas, Mexico

4 University of Michigan, Ann Arbor, MI, USA

5 La Via Campesina/ECOSUR, San Cristóbal de Las Casas, Mexico

6 Stanford University, Stanford, CA, USA

Introduction

Agrarian scholars have long debated how political and economic processes mediate the production, circulation, and employment of agricultural knowledge (Cleaver 1972; Dissanayake 1992; Morgan and Murdoch 2000). Central questions include: whose knowledge counts in advancing agricultural development (Thompson and Scoones 1994; Carolan 2006; Lewontin and Levins 2007; Ingram 2008)?; how do financial incentives privilege particular forms of agricultural knowledge production (Shepherd 2005; Kloppenburg 1988; Henke 2008)?; how do class, racial and other forms of power relations mediate access to knowledge (Gray et al. 1997; Daniel 2013)?; and what is the relation between knowledge and transitions to sustainable agriculture (Eshuis and Stuiver 2005; Stuiver et al. 2004; Caron et al. 2014)? Despite these extensive debates, agrarian studies scholars have devoted relatively little attention to analyzing how education contributes to food sovereignty.[1] This general absence is surprising given that agrarian movements and food justice organizations are deeply committed to educating for food sovereignty. From the landless movement in Brazil (Caldart 2004; Tarlau 2015a; Meek 2015b), to indigenous groups in Mexico and peasant networks in Cuba (Holt-Gimenez 2006; Baronnet 2008), social movements have prioritized forms of education that advance food sovereignty. These educational opportunities range from political leadership training to formal degree programs at various educational levels.

Food sovereignty means very different things in disparate geographic contexts, making it difficult to provide a universal definition of the concept. Despite this ambiguity, scholars agree that food sovereignty is a rights-based approach in which farmers, other producers, and communities are in control of their food system. Although La Via Campesina, a grassroots social movement, first articulated the concept, state-society interactions are increasingly central in advancing food sovereignty (Schiavoni 2016). For example, approximately 15 countries have developed policies around food sovereignty to date (Godek 2015). Political economic processes—in the form of public policies and economic incentives—are frequently tied to the institutionalization and expansion of food sovereignty (Wittman 2015).

The objective of this article is to comparatively analyze the thematic similarities and differences between disparate movements' food sovereignty education initiatives. Our understanding of "education" is intentionally broad—we focus on both non-formal education opportunities as well as formal degree granting programs. We highlight how many of same broad processes which shape food sovereignty, including state–society interactions, institutionalization, expansion, and political economy are intertwined with *educating* for food sovereignty.

We draw on two emerging theoretical frameworks to compare six examples of grassroots food sovereignty education programs that are embedded in social movements and other civil society organizations. First, the political ecology of education (PEoE) draws attention to how the distribution of power and resources among interconnected political and cultural entities mediates pedagogical processes—from tacit to formal learning—and related knowledge systems at various interconnected scales (Meek 2015a, b, c; Bradley and Herrera 2016). When applied to agriculture, the PEoE lens helps illuminate how movements gain access to the political and economic resources necessary to educate for food sovereignty. It also highlights how conflicting power relations between different educational institutions and forms of agricultural knowledge structure particular changes in agricultural practices, and ultimately transformations in the landscape itself.

It's here that the PEoE framework intersects with critical food systems education (CFSE)—a related perspective that integrates insights from food justice, political agroecology, popular education, critical pedagogy, and food sovereignty with traditional food systems education (Meek and Tarlau 2015, 2016). Many traditional food systems education courses seek to help students develop an interdisciplinary *understanding* of the food system. In contrast, the objective of CFSE is to leverage the broader educational system and innovative pedagogical techniques so that students and educators can utilize food system knowledge and agroecological practices to systematically dismantle the structural and ideological elements of the corporate food regime and develop transgressive subjectivities (Sawyer 2004). Drawing upon a synthesis of these two perspectives, our analysis of six case studies ultimately finds that political and economic processes lead to the genesis, advancement, institutionalization, and replication of forms of education that advance food sovereignty, transforming individuals' agrarian subjectivities, land relations, and understanding of agriculture itself.

Methods

The idea for the article arose at the University of Michigan's June 2015 Food Sovereignty conference. An emergent theme from the conference was the importance of education in advancing food sovereignty. Building upon that insight, the authors—most of whom participated in the conference—agreed upon the need for an article synthesizing experiences with food sovereignty education from diverse contexts. We chose six cases from across the Americas, including two U.S. programs, and one each from Cuba, Mexico, Brazil, and Bolivia (see Table 1). These cases were selected because each of the authors was directly involved with at least one of the projects. Some of the contributors are long-time activist-educators that have helped to found and contribute to these various programs, while others are scholars who became familiar with the programs through in-depth research. Each contributor draws on primary data gathered through long-term qualitative research conducted at the field site,

[1] A review of the Journal of Peasant studies—one of the central journals in agrarian studies—shows only several articles that focus on the connections between food sovereignty and education, and these few have just been published in the last 2 years (Meek 2015a; Tarlau 2015a). Considering other fields, it is clear that the linkages between agroecology and education have received extensive attention, but not in the context of food sovereignty (Lieblein et al. 2007; Francis et al. 2013; Hilimire et al. 2014).

Table 1 Summary of six case studies

Program	Launch	Country	Setting	Founders	Students
Colonia Pirai Agroecological Technical Institute (CPATI)	1970s	Bolivia	Rural	Teachers and doctors	College students
Landless Workers Movement (MST)	1980s	Brazil	Rural	Farmer–activists	Activists and students of all ages
Campesino a Campesina (MACaC)	Late 1990s	Cuba	Rural	Farmer–activists	Activists
FW Youth Development Program (FW)	2000s	USA	Urban	Teachers	Children and youth
Laboratorios para la Vida (LabVida)	2000s	Mexico	Rural and urban	Agroecologists and anthropologists	Teachers
East Bay Urban Farmer Field School (EBUFFS)	2010s	USA	Urban	Activists and urban farmers	Activist and urban farmers

including interviews, observations, document review, and historical research.[2]

The methodology for conducting the comparisons and writing the article took place over the course of a year. Rather than drawing upon a particular cross-case analytical methodology (Yin 2013), we chose to develop our analysis through an organic process of dialog. Our group came together four times in online group conference calls to develop and comparatively analyze the case studies. We first began with an initial discussion about the six cases. We transcribed our first conversation, and then read through the transcription and initial case study descriptions, coding for the common themes that cut across the cases. We did this coding informally (i.e. without data analysis software), and the themes emerged inductively. To reach our research objective, we focused on coding thematic areas of similarity and difference in how movements educate for food sovereignty.[3] We then worked both independently and in pairs to write long-versions of each of these case studies. We circulated these case study narratives within our group for review; each individual was tasked to independently highlight emergent themes that cut across the case studies. Within a subsequent discussion, we collectively compiled a list of the similarities and differences in each of the food sovereignty programs. As part of this process, we consolidated themes that were similar. Through these discussions we identified three overarching themes and three subthemes (see Table 2). We then worked independently and in pairs to produce both short versions of each case study, as well as the synthetic thematic results sections. The final two conference calls involved fine-tuning these analyses.

The six cases enable comparison on two levels. First, they take place in diverse educational environments. Whereas some case studies consist of farmer-to-farmer conversational learning, others involve more formalized classroom instruction, highlighting the varied nature of food sovereignty pedagogy. Second, we chose case studies from diverse geographic contexts, underscoring the transformative educational work that is being done in urban and rural areas and in several countries. In our cross-case analysis, we compare how these educational environments and geographical locations motivated the founders to start each of these programs, their pedagogical practices, and each program's financial sovereignty, expansion efforts, and relationships to the state (see a summary of this analysis in Table 2). These variables emerged as common themes during our iterative discussions. In our cross-case analysis, we analyze the common patterns and differences among these variables.

Case studies

The contexts, timeframe, types of founders and students, and the scope of each of the six case studies are considerably different. They were launched in both urban and rural settings across the Americas, as early as the 1970s and as recently as 2013. Teachers, educators and activists are both the founders and the students of many of these programs, as are youth, doctors, agroecologists and anthropologists (See summary in Table 1).

The oldest program we highlight, *Colonia Pirai Agroecological Technical Institute* (CPATI), was established by teachers and doctors in the 1970s in rural Bolivia. The program started as an orphanage, which later opened a grade school, and eventually, transformed into a three-year, residential agricultural technical college that enrolls between

[2] There is not enough space in this article to outline how each researcher collected and analyzed data, but for more information about some of these research methods please see Meek (2015a, b) and Tarlau (2015a).

[3] Some of the authors had already published research on the themes that had emerged, such as political economy (Meek 2015a), and horizontal pedagogy (Rosset 2015a).

Table 2 Cross-case analysis summary

Program	Issues that motivated the program's launch	Pedagogy models and content	Non-state finances	Expansion	Relations with the state
Colonia Pirai Agro-ecological Technical Institute (CPATI)	– Survival of indigenous farmers – Volatile 2008 food prices and climate change – Neoliberal social and economic models	– On-farm field schools – Applied research in own community – Indigenous knowledge, gender equity, agroecology, decolonization – Small-scale agribusiness practices	– Foundations – Self-funding	– Not trying to expand beyond the Institute	– No connection to State (in past, seen as "anarchist" and "communist")
Landless Workers Movement (MST)	– Land inequity – Education that does not valorize rural culture	– On-farm field schools – Critical place-based learning – Applied research in own community – Freirian *dialogo de saberes* – Seed saving as act of resistance – Cooperative production – Marketing and processing – Research skills/the scientific process	– Self-funding – Church and other donors – Volunteers – UN	– Large-scale expansion across country	– State used ag export profits to invest in agrarian reform – State applies MST's education model – MST maintains "movement schools" without state support
Campesino a Campesina (CaC)	– Nationwide food crisis after US embargo	– Farmer research on own farms – Freirian *dialogo de saberes* – Seed saving as act of resistance – Agroecology	– Self-funding – External donors	– Large-scale expansion across country – Members promote CaC to other networks	– Policies support peasant agroecology, but state yearns return to industrial agriculture
Food Warriors Youth Development Program (FW)	– Narratives of African Americans and agriculture – Food insecurity in communities of color – Lack of African-centered education	– African origins of food – Critical place-based learning – Cooking, nutrition, indigenous plant-based medicine, agroecology – Marketing and processing – Critical placed-based	– Foundations	– Difficulties expanding to more sites, schools – Creating handbook	– School where FW founders emerged was public charter – Today, only link is USDA
East Bay Urban Farmer Field School (EBUFFS)	– Devaluation/exclusion of communities of color – Food insecurity in communities of color – Agriculture education that marginalized social justice	– On-farm field schools – Farmer research on own farms – Horizontal knowledge production – Seed saving as act of resistance	– Foundations – Academic volunteers	– Not trying to expand field schools – Establishing regular participants	– No connection to STATE (formed because of little relevance of State funded training)

Table 2 (continued)

Program	Issues that motivated the program's launch	Pedagogy models and content	Non-state finances	Expansion	Relations with the state
Laboratorios para la Vida (LabVida)	– Food insecurity, health crisis, and erosion of indigenous livelihoods and traditions – Education divorced from local knowledge, traditions	– Formalized student research – Horizontal knowledge production – Research skills/the scientific process – Agroecology – Critical food system analysis	– Foundations – Volunteers	– Creating handbook – Launched network of state and international garden programs to share support, materials	– Program based in federal research centers – Most participating teachers in public schools

120 and 150 students. Students who attend are all on scholarships, recruited from indigenous communities from throughout Bolivia.

The second oldest program is comprised of diverse educational initiatives of the Brazil Landless Workers' Movement (*O Movimento dos Trabalhadores Rurais Sem Terra* or MST), a large national and grassroots movement struggling for agrarian reform through the occupation of unproductive land estates (described below). Since the 1980s, the MST has mobilized for access to and control over a range of educational programs, from infant education to universities, to train students of all ages and social movement activists—all with the purpose of contributing to the struggle for food sovereignty (Caldart 2004).

The third case study, the *Campesino a Campesino Agroecology Movement* (MACaC; Farmer-to-Farmer Agroecology Movement), is also embedded within a grassroots social movement in Cuba, the National Association of Small Farmers (ANAP). Since MACaC began in the late 1990s, it has helped facilitate the agroecological transformation of roughly half of all peasant farms in Cuba through its horizontal peasant pedagogy.

Two programs began in the 2000s. Laboratorios para la Vida (LabVida) is an action research program that was launched in Southern Mexico to train teachers from elementary, middle and high school (up to 18 years of age) and college professors on agroecological production and local foodways. Its founders are agroecologists and an anthropologist based at Mexican public research centers in San Cristóbal de Las Casas, in the state of Chiapas. After realizing that working directly with schoolchildren would limit the reach of the program, LabVida began training teachers to establish school gardens and use them to strengthen formal education. An initial series of six, 4-hour workshops evolved into a 120-hour certificate program.

The other program that emerged in the 2000s is the Detroit, Michigan *Food Warriors Youth Development Program* (FW). FW grew out of a food security and community gardening project run by educators at the Nsoroma Institute, one of Detroit's African-centered charter schools at the time. These educators later went on to co-found the Black Community Food Security Network, where FW became one of several food-focused youth internship and summer programs. Today, FW serves over 50 children in elementary and middle schools (up to 14 years of age) through three afterschool programs.

Finally, urban farmers and activists launched the East Bay Urban Farmer Field School (EBUFFS) in 2013 to engage in co-learning with other farmers and activists in the East Bay of the San Francisco Bay Area. The activists involved are diverse, working on issues from unemployment to the school-to-jail pipeline, public education, social work case management and food justice. Inspired by the CaC methodology, this urban agriculture education group promotes agroecological farming and creates learning opportunities through visits to different farms that focus on technical skills development, community building, and leadership development.

Program origins

Each of these diverse programs converged on food sovereignty education as a response to three issues spurred by industrialized food systems: social inequality; food insecurity; and decontextualized education systems. First, the discrimination and inequality that many of the actors responded to was tied to structural marginalization that black, brown, peasant, and indigenous communities were suffering and the disregard for their agricultural and culinary knowledge. The MST, the largest and most influential agrarian reform movement in Latin America (Branford and Rocha 2002; Wright and Wolford 2003; Carter 2015), is the most obvious example. The MST arose outs of a concern for societal and economic inequities surrounding land distribution. The MST's food sovereignty education programs emerged from a recognition that agricultural knowledge, like land, is highly concentrated in the hands of elites, and inaccessible to marginalized groups. The MST seeks to "break down the fences around knowledge," in addition to those surrounding land, valuing alternative forms of agricultural knowledge and practice (Diniz-Pereira 2005; Barbosa 2015, 2016).

Many of the other programs we feature were also motivated by stories of discrimination and inequities. In Bolivia, the CPATI founders began their original grade school concerned about the discrimination and poor quality of education their orphans were receiving in neighboring rural schools. Later, when the institution transitioned to offer post-secondary training in agroecology, a concern for the survival of indigenous farmers motivated them to recruit and guarantee scholarships for indigenous youth from across Bolivia. In Chiapas, Mexico, too, LabVida formed in part as a reaction to the rapid urbanization and industrialization of farming and food systems, which is eroding rural livelihoods and indigenous agroecological traditions.

The two US, urban programs have common roots in the discrimination communities of color were experiencing in the food system. In Detroit, FW challenges the common narrative that people of African descent are only historically connected to agriculture through enslavement and sharecropping. The Black Community Food Security Network and FW rose out of a desire to ensure that African American histories and communities not be "marginalization and maligned", according to Hanifa Adjuman, a co-founder of FW and DBCFSN. In part, these co-founders were responding to a concern that young, white suburbanites moving into the city were disregarding the rich legacy of urban agriculture in Detroit (White 2011). Local food justice activists who formed EBUFFS in Oakland, CA similarly observed the systematic devaluation of communities of color over the course of many decades (Cadji and Alkon 2014; McClintock 2011; Nelson 2011), and the race- and class-based exclusiveness of prominent sustainable agriculture networks in the region (Guthman 2008).

Persistent food insecurity is the second common issue that motivated the formation of all six programs, although the causes of food insecurity varied. EBUFFS activists and FW educators observed the concentrated food insecurity in communities of color and saw jobs in urban agriculture as a key to not only improving access to food but also to escaping poverty and increasing local control of the food system (White 2010; Sbicca 2012). In Cuba, after the collapse of the socialist bloc in Europe in 1989 and the United States trade embargo, the country was no longer able to import sufficient food, or the machinery, inputs and petroleum needed to maintain a conventional production model. The economic and food crisis that ensued in the 1990s spurred ANAP to seek alternative approaches to recover and boost national food production. More recently in Bolivia, the CPATI coordinators began focusing on food sovereignty and agroecology after they experienced volatile food prices during the 2008 global food crisis. This coincided with observations that industrialized practices were causing extreme soil degradation and that climate change was affecting the survival of rural communities and indigenous farmers. Similarly in Chiapas, LabVida coordinators were concerned about indigenous communities of Mayan descent that consistently rank at or near the bottom of Mexican states for human development indicators and face persistent food insecurity. They saw how globalization of the food system was causing price volatility, increased reliance on purchased, processed foods, and a crisis in diet-related disease that rivals the United States.

Finally, a de-contextualized education system was the third, interconnected concern that motivated action across all these cases. In EBUFFS, some members did not see the relevance of the technical assistance cooperative extension agents and sustainable agriculture educators offered. Looking for training that honored, rather than marginalized the social justice elements of their work, the farmers invited other urban farmers and food justice activists to their farm to learn and teach, initiating EBUFFS' farmer-to-farmer model. In Detroit, the K-8 Nsoroma Institute—the African-centered education program out of which FW emerged—was part of a broader movement among African American leaders to establish schools based on Nguzo Saba, the Seven Principles of Kwanza and a collective worldview. According to Adjuman, the school's focus on food was based on the sense that "food is a central part of what we are as a people…the foundation of our relationship to agriculture as African descendants".

In Chiapas, LabVida arose in response to an educational system that relies on a national curriculum, largely divorced from local knowledge and traditions (Klein et al. 2014). Teachers in state-run schools tend to be from cities, do not speak local languages, change schools frequently and rarely live in the communities where they work. Mexican public education prepares children and young adults in rural areas for urban jobs. By attending school, children lose much of their connection with the land and the knowledge of how to work it. At school, they often eat store-bought, processed food. Children in the city develop little understanding or appreciation of the rich cultures of the surrounding countryside. The garden work LabVida offers, in contrast, builds connections between school and community, providing experiential, culturally relevant learning opportunities for rural and urban students.

Similarly, in Brazil, MST founders also realized that traditional rural schools do not value rural culture, and seek to prepare rural students for life in urban centers. The MST sought to address this problem by developing alternative pedagogies for rural, public schools, based on local traditions, collective work practices, peasant culture, and student self-governance (Barbosa 2015, 2016). Also during the 1980s in Bolivia, the CPATI instructors saw themselves as contributing to a larger social and economic revolution during periods of dictatorships and the introduction of neoliberal economic models. The emphasis on agroecology today challenges the paradigms of industrialized agricultural education; it "encourages the restoration of local food, knowledge and biodiversity…[and changes the] model of accumulation towards one of redistribution with the active participation of all farming families and consumers" (CPATI 2015).

Pedagogical practices

Educational content

In educating for food sovereignty, the six cases we analyzed incorporate a wide variety of content. All groups emphasize agroecological principles—teaching why they matter and how to implement them. However, as their understanding of agroecology is political (Molina 2013), the education emphasizes the economic, political, and cultural dimensions of the agroecological transition (Rosset 2015; McCune et al. 2016). This ranges from framing seed saving and agroecology as political acts of resistance (MST, EBUFFS, MACaC) to deliberately teaching and linking a food sovereignty framework with indigenous knowledge, decolonization, and gender equity (CPATI, FW, EBUFFS). In the case of the MST, activists have drawn on the writings of Soviet pedagogues Anton Makarenko and Moisey Pistrak to strengthen the movement's organizing capacity and enhance community self-sufficiency by teaching communal social practices and cooperative production strategies (Tarlau 2012; Rosset and Martinez-Torres 2012). Several groups also provide opportunities for members to learn about marketing and value-added processes, such as cheese making (MST, FW, CPATI). To cultivate an ethic of self-care and cultural pride, some groups teach cooking, nutrition, and/or indigenous plant-based medicine (FW and LabVida), and FW also teaches about the African origins of food and agriculture.

While every group teaches a unique combination of these topics, the educational content reflects each groups' unique goals, capacities, and place. Political ecologists emphasize how ecological change is interrelated with larger historical and political economic processes; similarly, we see here that across these case studies the diversity in content is in part a function of the unique historical, political, cultural, economic, and ecological contexts in which these groups operate. Yet, variations in content also helps broaden our understanding of critical food systems education, because each group's pedagogical content is a function of their understanding of what constitutes food sovereignty, and how education can serve to mediate the opportunities and constraints towards advancing it. The content of critical food systems education will look dramatically different in divergent cultural contexts.

Educational approach

All of the case studies are theoretically grounded in the tradition of popular education, particularly Paulo Freire's work on critical pedagogy, dialogic education, and agricultural extension (1973a, b, 1986). These movements and organizations apply this approach by emphasizing education as a tool for developing critical consciousness and encouraging students to learn from their own reality, recognizing the power structures that shape their food environment, and focus on transforming the social and economic inequities in their communities.

Each of the groups prioritizes some form of experiential training as a form of critical pedagogy. Hands-on lessons occur on local farms, and topics are chosen that are most likely to have an immediate or short-term impact in each place, such as agroecological techniques for pasture rejuvenation (MST) or greenhouse irrigation (EBUFFS). Farmer research is another important experiential learning tool, although each initiative uses it differently. In some contexts, the research is informal or incidental. For example, in the Cuban and EBUFFS context, farmers are encouraged to experiment and innovate on their own farms, and learn from their own experiments. Additionally, as a newly formed group, EBUFFS participants consider the entirety of their collective work as experimental. By contrast, more formalized student research is integral to the educational initiatives of LabVida, the MST, and CPATI. In these cases, teachers encourage students to use participatory action research to scientifically answer questions stemming from critical pedagogy. In the MST, CPATI and LabVida, students conduct participatory action research as part of an applied field project back in their home communities. Whereas in CPATI the objective is to make their agroecological education more visible and to spread concepts they learned in school to their rural communities, in the MST and LabVida it is also a method for developing critical consciousness. Students in MST programs engage in critical place-based learning (Gruenewald 2003), researching the tension between hegemonic and counter-hegemonic forms of agricultural production in their communities, and socializing their results as a means to cultivate broader mobilization. With participatory action research tools, LabVida students reflect on their own teaching practice. FW engages youth in similar critical analysis and reflection, and has had participating youth engage other youth and adults in the same exercises. One such lesson, "Who tells you what to eat?," guides students through a process of critically deconstructing media messages and then creating their own, using jingles and slogans that counter food marketing messages they critique.

Experiential learning opportunities for individual and/or collective reflection are also frequently tied directly to foodways in many of the programs. In at least four of the cases (FW, LabVida, EBUFFS MST), meals or tastings bring people together for reflection, which includes charting personal goals, evaluating group work, responding to local stressors, and honoring local culture.

The groups also share a similar vision of participatory knowledge production. Dialogic education, whereby knowledge is produced among equals through communication (Freire 1986), is an explicitly horizontal form of knowledge production that breaks down the dichotomy between teachers and learners. In this context, everyone has something to share and a perspective that is valuable. Horizontal knowledge production is exemplified by the Campesino-a-Campesino (CaC) approach, which the Cuban MACaC, EBUFFS, and LabVida cases utilize to varying degrees. The CaC approach involves farmer-promoters, who have either innovated new solutions or rediscovered older traditional solutions to common agricultural problems, using a popular education methodology to share them with their peers. For example, in EBUFFS, an urban farmer with experience installing an irrigation system on his own farm visited an urban farm needing to install such a system, guiding a discussion of how the system could work in a new setting. LabVida's constructivist approach encourages teachers to put local knowledge into horizontal dialog with the formal state curriculum, allowing students to learn from each other and teachers to learn from students and the surrounding community. In the MST, and La Via Campesina more broadly, horizontal knowledge production also takes the form of *diálogo de saberes*, which is translated as a "dialogue among different knowledges and ways of knowing" (Rosset and Martinez-Torres 2012, p. 2). This Freirian pedagogy seeks to bring peasants and extension agents into dialogue, with the goal of synthesizing traditional peasant forms of knowledge and academic forms to forge new knowledges, practices, and social relations (Leff 2004; Meek and Simonian 2016).

All of these horizontal methods are based in a critique of traditional approaches to agricultural extension, where knowledge is produced by experts and then imparted to what are perceived of as "backwards" and knowledge deficient farmers. This conventional vision of agricultural extension has been closely linked to the Green Revolution and the advance of capitalism in rural areas (Cleaver 1972). From a political ecology perspective, all cases can be seen as transforming intertwined material and immaterial territories. For Brazilian agrarian geographer Bernardo Mançano Fernandes, material territory consists of landforms and infrastructure, whereas immaterial territory is the ideologies associated with particular landforms (Fernandes 2009; Rosset and Martinez-Torres 2012). Education can structure the relationships between material and immaterial territories, because the dominant ideas of a society dictate norms concerning proper forms of agricultural management. As an example, traditional forms of agricultural extension based on a Green Revolution model of high inputs impart *technical* knowledge to create a *technical* landscape. In contrast, whether it singularly horizontal, or occurring in tandem with formalized classroom instruction, the learning strategies in all of these cases teach students to challenge the dominant means of production (Barbosa 2016), by developing agroecological practices, breaking reliance on fossil-fuel based inputs, and advancing new forms of sustainable production. These case studies illustrate the potential for critical food systems education to transform material territory in the form of higher farmer productivity. In the case of EBUFFS, the work is leading to greater productivity among urban farmers. Similarly, in Cuba, through MACaC the spread of agroecology in the peasant sector has coincided with a major relative increase in its contribution to national food sovereignty (Machín Sosa et al. 2010, 2013; Rosset et al. 2011).

Each of these case studies highlights the potential of education to affect levels of political participation and civic involvement in the food system. All cases pointed to ways they have motivated and equipped participants to become food movement leaders, or "paradigm shifters", as a FW coordinator put it. Students go on to engage their communities to change perceptions and practices (FW, LabVida, MCaC, MST), or become involved directly in regional and national food justice movements (EBUFFS). Graduates also become local municipal authorities, extending the potential reach of their impact on agriculture policy and development (CPATI; MST).

Expansion and institutionalization

In addition to educational content and approach, educational scholars have shown that the political economy of education, or the influence of public policies and financial incentives (Carnoy 1985; Heyneman 2005), is key to understanding how market-based ideologies are promulgated in schools. Yet, from a political ecology of education perspective these questions of political economy are equally important to understand how subaltern forms of agricultural knowledge and practice are promulgated (Meek 2015a, b). We found that the extent to which each of the cases is able to disseminate and institutionalize its programs is related in complex ways to questions of financial sovereignty, and state influence.

Financial sovereignty

We identify funding as a central mechanism behind the expansion and institutionalization of critical food systems education across these case studies. Three main sources of funding—external non-governmental funding, state funding, and self-financing through productive activities—enabled these programs to grow and support teacher training, student travel, curriculum development, the construction of experimental gardens, and extra-curricular activities.

Both of the U.S.-based food programs have primarily relied on external funding, in the form of governmental grants and funding from foundations. FW, which is an educational program institutionally located within the Detroit Black Community Food Security Network, has received funding from the Kellogg Foundation and the US Department of Agriculture. This latter grant was part of a multi-institution 5-year grant, the Detroit Child Health Incubator Project, to support its work with children ages 2–8. While this large state grant provided the needed capital and materials, it has also limited FW's ability to reach older youth in the program (9–13 year olds). For these reasons, the program has searched for ways to self-fund as well, through youth farm stands, but the organization continues to primarily rely on national government or foundation funding. Similarly, EBUFFS in California has relied on external funding, more specifically resources from the Heller Foundation. Other non-profit backers, however, including Food First, Urban Tilth, and the Center for Popular Research, Education, and Policy, hope that EBUFFS will grow increasingly autonomous.

In Latin America, in contrast to the U.S., there is a tendency to rely more on self-funding and the creative use of state resources (rather than one-time grants). For example, in Bolivia, CPATI always strove to become financially autonomous. It launched pork, dairy, chicken and egg, and livestock feed operations to teach its orphans, and later college students, self-sufficiency and agricultural skills, while also contributing significantly to the school's expenses and scholarships. Although the school has also received grants from outside sources, such as donors in Spain and the Inter-American Development Bank, a 2015 report by school administrators cites that the organization is now close to financial independence (CPATI 2015).

In contrast to CPATI, the LabVida teacher-training program in Mexico initially mobilized resources through an outside grant from the Kellogg Foundation and from the public research centers that pay the salaries of the academics involved in the program. However, by training teachers in public schools to contribute to agroecological "literacy" and food sovereignty, LabVida leverages the tremendous resources invested in education by the Mexican State. Continued funding will likely rely on a mix of institutional resources, grants, and payments received for training and consulting. By providing little material support to participating schools, however, LabVida avoids creating dependency, a common pitfall for government programs. Instead, the program promotes autonomy and community by encouraging teachers to draw upon local resources and knowledge. Requesting help from families and community members to set up and maintain gardens also enriches the program, reduces the burden on teachers, and contributes to program continuity when teachers transfer schools.

In Cuba and Brazil, the educational programs we explore are both embedded in grassroots social movements, rather than a single school or institute. These programs, which affect tens of thousands of people, exemplify how a combination of self-funding through productive activities and the creative use of state resources can help institutionalize and expand radical educational programs. In the case of Cuba, external donors funded MACaC on a pilot basis in the late 1990s (Machín Sosa et al. 2013). Around 2000, when that funding began to dry up, the National Small Farmers' Association (ANAP) determined that this agroecological educational program should become a "movement" and rely on its own resources. The program expanded much more rapidly since this period, funded primarily through the self-imposed quota from the sales of member cooperatives (Machín Sosa et al. 2013).

The MST has negotiated financial autonomy and dependency as it has developed its educational programs. Similarly to the Cuba case, the MST's first educational initiatives in the early 1980s relied on external support from progressive priests in the Catholic Church and national allies who would donate educational resources to the occupied encampments, in the form of money, schooling materials, or pedagogical expertise. For example, in the early-1980s two members of the Paulo Freire's literacy team visited one of the MST's camps in Rio Grande do Sul to talk to teachers about how to integrate Freirean pedagogies into their teaching (Tarlau 2015a). The teachers in the camps who developed these programs were entirely volunteer—activists in the movement who had some background or training in education. In 1987, the MST created a National Education Sector to think more strategically about transforming Brazilian education to support small-farming and collective agricultural production in the countryside (Barbosa 2016).

The MST, much like the Mexican LabVida program, has also partnered with state governments in offering teacher-training programs that are grounded in an alternative educational approach. The movement has run large-scale literacy campaigns throughout Brazil, in partnership with state governments and university partners. These MST-state collaborations were possible because despite the MST's antagonistic relationship to the state, the MST's educational initiatives increased the state's capacity to provide educational access to citizens (Tarlau 2015a; Barbosa 2016). As the MST's educational initiatives gained more public recognition in the mid-1990s, international organizations such as UNESCO and UNICEF also began to fund these programs. Although the MST has found creative ways to use state and other resources to fund their educational initiatives in the formal school sphere, the MST's internal political education programs (administered independently from the state) are primarily funded from donations from the MST's agrarian reform settlements.

Taken together, these case studies highlight two important findings concerning financial sovereignty. First, all of our case studies received financial support at some point from external entities. This makes sense when seen through the sociological lens of resource mobilization theory (McCarthy and Zald 1977), which highlights how movement success frequently hinges upon creating material linkages with supporters. Similarly, drawing upon political opportunity theory (Tarrow 1994; McAdam and Snow 1997), we can better understand how movement reliance upon external funding is often contextual. For instance, the MST was able to get PRONERA approved in 1998 after the massacre of 19 MST activists in Eldorado dos Carajás 2 years previously, during which time there was intensive international pressure on the Brazilian government (Tarlau 2015a). However, our second finding on this theme is that developing financial autonomy is a process—one frequently intertwined with external reliance—and that movements and civil society groups seek to develop sovereignty over their educational funding through the commercialization of agricultural produce, and local donations (MACaC, CPATI, LabVida). When brought together, these two findings highlight how financial sovereignty is much like food sovereignty: not an end state, but rather an organic process that evolves as political opportunities emerge, sources for institutional partnerships develop, and the movement's base creates its own self-sustaining mechanisms.

Program expansion

The transfer and expansion of these educational initiatives differs drastically. The four educational programs that are located in a single school or program structure (FW, LabVida, EBUFFS, CPATI) all have had difficulties expanding, although as we will subsequently explore, in some cases teachers and activists are implementing creative solutions to increase the impact of these programs. In contrast, the two education programs embedded in social movements (Cuba, Brazil) have illustrated an impressive capacity for expansion and growth, although sometimes this can dilute the original aims of these educational programs.

The most recently developed program, EBUFFS in California, has not sought to expand its urban field schools, but rather to consolidate a group of regular participants from among the food justice and urban agriculture community. In Detroit, lack of funding has restricted FW program expansion, but instructors are developing a packet of activities to share their curriculum with other schools and teachers. Lessons are being aligned with state standards to facilitate their integration with teachers' daily practice. Similarly, in Mexico, LabVida has published a school garden handbook containing lesson plans designed and tested by teachers who participated in the certificate program and linked to Mexico's national curriculum. The handbook conveys the potential of school gardens for strengthening education across subject areas, helping students develop a scientific attitude, build critical consciousness around agro-food systems, and bolster school-community relationships. Although the push for educational standards is frequently associated with neoliberalism, and the transformation of public education into a private good (Apple 2006), the case of LabVida highlights how movements can creatively adapt to these situations—in which teacher autonomy is relatively restricted—and still implement some counter-hegemonic goals in schools. LabVida has also founded state and international school garden networks (redhuertos.org) that facilitate the exchange of experiences, materials, and inspiration among schools and other programs.

In Cuba, as external funding dried up, ANAP mobilized internal resources to train successive generations of agroecology promoters (Machín Sosa et al. 2010). Using its "mass mobilization" methodology, ANAP holds each of its members accountable for promoting the CaC movement within his or her sphere of responsibility. Much of their work is ideological, emphasizing the peasant sector's 'historic mission' to feed the Cuban population. This methodology has allowed agroecological production to expand much faster than the conventional extension strategies ANAP employed before it started using CaC (Rosset et al. 2011; Machín Sosa et al. 2013).

In Bolivia, CPATI has grown in both size and breadth, from working with less than two dozen orphans on a single 100 ha site in 1973, to teaching over 150 technical college students across three sites that sit on 385 ha today. But beyond its own Institute, it has never tried to actively integrate its methods into the public school or university system. In contrast in Brazil, the MST's educational initiatives expanded rapidly throughout the 1990s and 2000s. Between 1998 and 2011, PRONERA has offered 320 courses, ranging from literacy training and high-school to university degrees, involving 82 different educational institutions, and training 164,898 students living in agrarian reform areas (IPEA 2015). Agroecological courses offered through PRONERA include agronomy, agricultural extension, pedagogy, and forest management. In addition to PRONERA, the movement currently has 2000 schools in its encampments and settlements, with over 8000 teachers attending to 250,000 students (Carter and Carvalho 2015). However, the MST's ability to implement their educational proposal in these schools varies widely across the country (Barbosa 2016). The case of the MST is a particularly stark illustration of why it is critical to pay attention to the state when analyzing program expansion and transformation.

Certain cases suggest that critical food systems education can increase educational opportunities for populations that would not normally have access to K-12 or post-secondary schools. In Bolivia, indigenous youth that participate from around the country would likely not have an opportunity to receive a post-secondary education without the scholarships CPATI provides. Similarly, the MST's educational initiatives have also given hundreds of thousands of rural communities access to schools, childcare centers, literacy programs, adult education, and higher education courses. These cases illustrate that social movements and other civil society groups—which mobilized around concerns about structural discrimination—can play an important role in 'breaking down the fences around knowledge' and improving access to education.

Relationship to the state

Finally, these six educational programs have had radically different degrees of interaction and collaboration with city, state, and national governments. Some groups have rejected the involvement of the state in their educational programs, while others have identified the state as the primary actor responsible for providing these educational courses.

The two movement-based programs offer an interesting contrast to state-society relations. In the case of Cuba, the government is known internationally as being "pro-peasant," however in reality, the Ministry of Agriculture only accepted agroecological peasant agriculture "during the crisis" and now yearns for a return to large-scale industrial agriculture "once the embargo is lifted." Thus, while ANAP has won a large number of public policies that are favorable to peasant agroecology,[4] the main thrust of government policy is still geared towards facilitating a "return to the future" of industrial agriculture. Consequently, the core budgets of ANAP and MACaC have not come from the Cuban government, but rather from a self-imposed, voluntary internal tax on sales by member co-ops.

Similar to Cuba, the Brazilian government has also promoted industrial agricultural production and a primary export model, despite the election of the Workers' Party to power in 2003. Nonetheless, during the first three terms of the Workers' Party government, the federal government used profits from primary exports to invest more money in agrarian reform settlements. In the educational realm, the government embraced the MST's educational proposal as an approach for all rural populations, known nationally as *Educação do Campo* (Education of the Countryside) (Barbosa 2016). In 2004 the Workers' Party government created an office for *Educação do Campo* in the Ministry of Education with dozens of new programs (Tarlau 2015a).[5] The Workers' Party also gave more funding to PRONERA educational programs. However, although *Educação do Campo* is now nationally recognized, many MST activists fear that many of the more radical components, and the connection between the educational proposal and an alternative food system, have been lost. For example, agribusinesses are now even declaring "support" for *Educação do Campo* (Tarlau 2015b). Despite these contradictions, the MST continues to engage in a complex process of negotiation and contestation with different levels of the Brazilian state for educational resources and recognition of their schools throughout the country. Importantly, the MST *simultaneously invests in its own internal educational programs*, independent from the state, and has established dozens of *escolas da formação*

[4] These Cuban pro-peasant programs include: distribution of land, crop insurance, the National Programs for Urban Agriculture and Suburban, for Production of Biological Inputs, for Animal Traction, for Production of Organic Matter, the Forum Movement on Science and Technology, the Growing Popular Rice Program, the Participatory Plant Breeding Program, programs to acquire farm animals, to achieve decentralized self-sufficiency in dairy products, to redesign curricula, to produce special programs on TV and radio and newspaper coverage (Machín Sosa et al. 2013).

[5] Even before the PT took power, in 2001, the federal government passed national guidelines for *Educação do Campo*. This was due to a coalition of agrarian social movements and the rural union movement, who put pressure on the federal government to embrace these policies.

(political training schools) across the country that offer courses on agroecology and food sovereignty without the direct state involvement (Barbosa 2015, 2016).

In the four other cases, these state-society relations have been less complicated. In California, EBUFFS formed in response to the limited relevance of extension and other state-funded technical assistance programs to urban farmers, so their field schools have no connection to the state. In Detroit, before FW became a formal program, the founders of the program who began a school garden project had a direct relationship to the state when their school became a public charter (1997–2012). However, when the principal of the school co-founded the Detroit Black Community Food Security Network, FW was officially launched and hosted by the Network and since then, FW has only been connected to the state through national (USDA) government funding. In Bolivia, CPATI has maintained an autonomous relationship from the state, and in its early days was referred to as "an authentic island—with anarchist values" (CPATI 2015, p. 4). The founders of the school have generally maintained a good relationship to the state, although the school has also been critiqued for being a "breeding ground for communists" (CPATI 2015, p. 11).

Finally, in Mexico, although LabVida does not rely on state funding, the program attempts to mobilize state resources to support food sovereignty. The program and its research team are based at Mexican federal research centers (ECOSUR and CIESAS). Participating teachers are primarily from public schools associated with the Public Education Secretariat (SEP).[6] The certificate program is validated by ECOSUR but is not recognized by the SEP as part of teachers' continuing education requirements. Nonetheless, the program is designed to help teachers link school gardens to the specific elements of the official curriculum they teach. Recently, state education authorities began to support school gardens through the environmental education program *Educar con Responsabilidad Ambiental* (ERA). Anecdotal evidence suggests that while some teachers see ERA as an imposition, teachers in LabVida appreciate the help from ERA, suggesting that state support for school gardens could be key for their broader adoption.

Together, these cases suggest that if food sovereignty education programs are small-scale and local, a relationship with the state might not be necessary. In the case of the two U.S. programs, this autonomy from the state is purposeful, as the anti-racist nature of these two cases is most often antithetical to the U.S. curriculum. However, if educational initiatives develop a national scope, a relationship with the state is probably inevitable, and movements are then faced with the challenging task of wielding state power for counterhegemonic ends, without losing sight of their original goals. The only case where this movement-state relationship was not necessary, even with educational initiatives on a national scale, is the second iteration of the MACaC movement in Cuba when these programs became embedded in social movements and self-sufficient through internal funding.

Conclusion

This article helps to synthesize the political ecology of education and critical food systems education frameworks, by drawing attention to the relations between financial elements, educational approaches, and food sovereignty movements. Certain cases—CPATI and MCaC—suggest that food sovereignty can be a tool to advance financial sovereignty, and ultimately educational sovereignty. Educational sovereignty is a concept that has natural linkages with food sovereignty. Educational sovereignty centers upon communities' right to challenge enshrined systems of educational inequality, and develop their own educational systems (Moll 2002; Lomawaima and McCarty 2002; Moll and Ruiz 2005). By marketing agricultural products produced at the school (CPATI) and adding a tax on agricultural sales (MACaC), certain social organizations and movements showed how local control over the food system can contribute to financial control over an educational program, enabling alternative forms of pedagogy directed towards food sovereignty. Importantly, these examples highlight that both social movements and civil society organizations can become financially sovereign—in other words, this is not merely a capacity of solitary organization. These examples contribute to our understanding of how to actualize critical food systems education's primary objective of educating for food sovereignty, which means that pedagogy contributes to a community's physical control over food system management (Meek and Tarlau 2016, p. 252).

Yet, these case studies also pointed to the fraught interrelationships between financial sovereignty and educating for food sovereignty. If obtaining financial resources requires becoming beholden to an external entity—whether it is the state or a donor—the original transformative intent of the movement can be diluted. Following this logic, institutionalization can be the death knell for critical food systems education. Yet, as this comparative study showed, in certain cases it is *through* institutionalization that movements preserve their ability to advance emancipatory objectives. As we discussed, the MST, for example, has helped create the National Program for Education in Agrarian Reform (PRONERA), which funds many MST critical agroecology courses in partnership with state

[6] Others teachers are from private schools recognized by the SEP, and a few teach at autonomous schools.

and federal universities and technical institutes (Barbosa 2016). However, these results also highlight the incompatibility between some civil society organizations and the state (EBUFFS, FW)—when groups are struggling against structural discrimination, institutionalizing within the state may antithetical to a broader political project. Such results push both political ecologists of education and critical food systems educators to think in more nuanced ways about the relations between political economic processes, civil society, and the state. In certain contexts, to gain autonomy requires careful and sustained work within state apparatuses—and in others, the context precludes particular forms of engagement.

The dialogue between the case studies, which take place in both rural and urban contexts, calls for deeper analyses of the pedagogical relations between these spaces. Food sovereignty—historically a rural political project (Wittman 2011) —has increasingly become a focus of urban movements. As assumptions of a strict rural–urban divide continue to be deconstructed (Lerner and Eakin 2011), it becomes particularly important for scholars and activist to look at the pedagogical cross-fertilization that is taking place between the Global South and North, and rural and urban movements. The interrelations between the EBUFF and MCaC case studies are exemplary of this trend; as we showed, EBUFFs adapted a Cuban horizontal peasant pedagogy to an urban US context.

Similarly, the article pushes scholars, activists, and educators to rethink the horizon of critical food systems education. Guthman (2008) has used the phrase 'politics of the possible' to refers to one's perceived repertoire of viable tactics and visions for political economic and social change. The politics of the possible within many traditional forms of food systems education, such as garden-based learning programs, is largely limited to shaping the purchasing choices of future white middle-upper class consumers (Meek and Tarlau 2015, 2016). This reformist ideal does not address systemic class and race-based inequity in the food system. As these case studies demonstrated, critical food systems has the potential to create agents of change, increase agricultural productivity, and improve educational access.

At a broader level, these cases provide a unified response to Guthman's query about the politics of the possible, illustrating that "Another Type of School System" is possible, which represents a counter-proposal to the dominant capitalist approach to schooling and agricultural knowledge production. This represents what Erik Olin-Wright (2012, p. 9) refers to as a "real utopia," which allows us to "envision the counters of an alternative social world that embodies emancipatory ideals." Each of these cases constitutes in different ways real utopias, and many have already inspired hundreds of organizations and activists to re-think their educational practice.

As these case studies demonstrate, popular education is playing a progressive role in transforming food systems. We are in the midst of a rapidly changing political climate. As this article demonstrated, political opportunities are powerful moments for innovation in critical food systems education. Between the recent ouster of Brazilian President Dilma Rouseff, the opening of diplomatic relations between Cuba and the United States, and the upcoming elections in Bolivia, it is clear that scholars and activists need to be vigilant in ensuring that important gains in critical food systems education—such as PRONERA—are not lost. In this rapidly evolving political moment, ongoing critical analysis is needed of the role that education plays in either perpetuating or contesting the contemporary food system, both in formal and informal education systems.

Acknowledgements Each of the authors would like to acknowledge the contributions of the farmers and activists with which we worked. Bradley's work with EBUFFS was financially supported by the AFRI project: Food Dignity.

References

Apple, M.W. 2006. *Educating the "right" way: Markets, standards, god, and inequality*. New York: Routledge.

Barbosa, L.P. 2015. *Educación, resistencia y movimientos sociales: La práxis educativo-política de los Sin Tierra y de los Zapatistas*. Mexico City: LIBRUNAM.

Barbosa, L.P. 2016. Educação do campo [Education for and by the countryside] as a political project in the context of the struggle for land in Brazil. *The Journal of Peasant Studies*. https://doi.org/10.1080/03066150.2015.1119120.

Baronnet, B. 2008. Rebel youth and Zapatista autonomous education. *Latin American Perspectives* 35 (4): 112–124.

Bradley, K., and H. Herrera. 2016. Decolonizing food justice: Naming, resisting, and researching colonizing forces in the movement. *Antipode* 48(1): 97-114.

Branford, S., and J. Rocha. 2002. *Cutting the wire: The story of the Landless Movement in Brazil*. London: Latin America Bureau.

Cadji, J., and A. Alkon. 2014. One day, the white people are going to want these houses again. In *Incomplete streets: Processes, practices and possibilities*, eds. S. Zavestoski, and J. Agyeman, 154–175. London: Routledge.

Caldart, R.S. 2004. *Pedagogia do Movimento Sem Terra*. Sao Paulo: Expresso Popular.

Carnoy, M. 1985. The political economy of education. *International Social Science Journal* 37 (2): 157–173.

Carolan, M.S. 2006. Social change and the adoption and adaptation of knowledge claims: Whose truth do you trust in regard to sustainable agriculture? *Agriculture and Human Values* 23 (3): 325–339.

Caron, P., et al. 2014. Making transition towards ecological intensification of agriculture a reality: The gaps in and the role of scientific knowledge. *Current Opinion in Environmental Sustainability* 8: 44–52.

Carter, M., ed. 2015. *Challenging social inequality: The Landless Rural Worker's Movement and agrarian reform in Brazil*. Chapel Hill: Duke University Press.

Carter, M., and H.M. Carvalho. 2015. The struggle on the land: Source of growth, innovation, and constant challenge for the MST. In *Challenging social inequality: The Landless Rural Worker's*

Movement and agrarian reform in Brazil, ed. M. S. Carter, Chapel Hill: Duke University Press.
Cleaver, H.M.J. 1972. The contradictions of the Green Revolution. *The American Economic Review* 62 (1/2): 177–186.
CPATI (Colonia Pirai Agroecological Technical Institute). 2015. *Cuarenta Años de la Fundacion Colonia Pirai: Memoria Historica*. Santa Cruz: CPATI (Colonia Pirai Agroecological Technical Institute.
Daniel, P. 2013. *Dispossession: Discrimination against African American farmers in the age of civil rights*. Chapel Hill: University of North Carolina Press.
Diniz-Pereira, J.E. 2005. Teacher education for social transformation and its links to progressive social movement: The case of the landless workers movement in Brazil. *Journal for Critical Education Policy studies* 3 (2): 92–123.
Dissanayake, W. 1992. Knowledge, culture, and power: Some theoretical issues related to the agricultural knowledge and information system framework. *Knowledge, Technology & Policy* 5 (1): 65–76.
Eshuis, J., and M. Stuiver. 2005. Learning in context through conflict and alignment: Farmers and scientists in search of sustainable agriculture. *Agriculture and Human Values* 22 (2): 137–148.
Fernandes, B.M. 2009. Sobre a tipologia de territórios. In *Territórios e territorialidades: Teoria, processos e conflitos*, eds. M. A. Saquet, and E. S. Sposito, 197–215. Sao Paulo: Expressão Popular.
Francis, C., T.A. Breland, E. Ostergaard, G. Lieblein, and S. Morse. 2013. Phenomenon-based learning in agroecology: A prerequisite for transdisciplinarity and responsible action. *Agroecology and Sustainable Food Systems* 37: 60–75.
Freire, P. 1973a. *Pedagogy of the oppressed*. New York: Continuum.
Freire, P. 1973b. *Education for critical consciousness*. New York: Seabury Press.
Freire, P., et al. 1986. *Pedagogia: Diálogo e conflito*. Sao Paulo: Cortez.
Godek, W. 2015. Challenges for food sovereignty policy making: The case of Nicaragua's Law 693. *Third World Quarterly* 36 (3): 526–543.
Gray, I., et al. 1997. Power, interests and the extension of sustainable agriculture. *Sociologia ruralis* 37 (1): 97–113.
Gruenewald, D.A. 2003. The best of both worlds: A critical pedagogy of place. *Educational Researcher* 32 (4): 3–12.
Guthman, J. 2008. Neoliberalism and the making of food politics in California. *Geoforum* 39: 1171–1183.
Henke, C.R. 2008. *Cultivating science, harvesting power: Science and industrial agriculture in California*. Cambridge: The MIT Press.
Heyneman, S.P. 2005. The history and problems in the making of education policy at the World Bank. In *Global trends in educational policy*, vol. 6, eds. A.W. Wiseman, and D.P. Baker. San Diego: Elsevier.
Hilimire, K., S. Gillon, B.C. McLaughlin, B.M. Dowd, and K. Monsen. 2014. Food for thought: Developing curricula for sustainable food systems education programs. *Agroecology and Sustainable Food Systems* 38: 722–743.
Holt-Gimenez, E. 2006. *Campesino a Campesino: Voices from Latin America's farmer-to-farmer movement for sustainable agriculture*. Oakland, CA: Food First Books.
Ingram, J. 2008. Agronomist–farmer knowledge encounters: An analysis of knowledge exchange in the context of best management practices in England. *Agriculture and Human Values* 25 (3): 405–418.
IPEA. 2015. *Relatório da II Pesquisa Nacional sobre a Educação na Reforma Agrária*. Brasilia: Instituto de Pesquisa Aplicada.
Klein, M. G., J.A. Girón, B. Hernández Zavaleta, M. López López, S.I. Navarro Martínez, B.C. Salazar Narvaéz, A. Saldívar Moreno, N.E. Santos Baca. 2014. Desde Chiapas: Criterios básicos para la exigencia de pertinencia cultural y lingüística en la educación, In *Desenmascarar la discriminación: La violencia del sistema educativo mexicano hacia los pueblos originarios y las personas con discapacidad*. eds. P.T. Baltazar, 69–94. San Cristóbal de Las Casas, Mexico: Incidencia Civil en la Educación.
Kloppenburg, J.R. 1988. *First the seed: The political economy of plant biotechnology, 1492–2000*. Cambridge: Cambridge University Press.
Leff, E. 2004. Racionalidad ambiental y diálogo de saberes. Significancia y sentido en la construcción de un futuro sustentable. *Polis. Revista Latinoamericana* 7: 1–29.
Lerner, A.M., and H. Eakin. 2011. An obsolete dichotomy? Rethinking the rural-urban interface in terms of food security and production in the global south. *Geographical Journal* 177: 311–320.
Lewontin, R., and R. Levins. 2007. *Biology Under the Influence: Dialectical Essays on Ecology, Agriculture, and Health*. New York: Monthly Review Press.
Lieblein, G., T.A. Breland, E. Ostergard, L. Salomonsson, C. Francis. 2007. Educational perspectives in agroecology: Steps on a dual learning ladder toward responsible action. NACTA Journal 51: 37–44.
Lomawaima, K.T., and T.L. McCarty. 2002. When tribal sovereignty challenges democracy: American Indian education and the democratic ideal. *American Educational Research Journal* 39 (2): 279–305.
Machín Sosa, B., A.M. Roque Jaime, D.R. Ávila Lozano, and P.M. Rosset. 2010. Revolución agroecológica: el movimiento de campesino a campesino de la ANAP en Cuba. *Cuando el campesino ve, hace fe*. La Habana: ANAP y La Vía Campesina. http://www.viacampesina.org/downloads/pdf/sp/2010-04-14-rev-agro.pdf
Machín Sosa, B., et al. 2013. *Agroecological revolution: The farmer-to-farmer movement of the ANAP in Cuba* [online]. Havana and Jakarta: ANAP and La Vía Campesina. http://viacampesina.org/downloads/pdf/en/Agroecological-revolution-ENGLISH.pdf. Accessed 21 May 2016.
McAdam, D., and D.A. Snow. 1997. *Social movements: Readings on their emergence, mobilization, and dynamics*. Los Angeles: Roxbury Publishing Co.
McCarthy, J., and M. Zald. 1977. Resource mobilization and social movements: A partial theory. *American Journal of Sociology* 82 (6): 1212–1241.
McClintock, N. 2011. From industrial garden to food desert: Demarcated devaluation in the flatlands of Oakland, California. In *Cultivating food justice: Race, class, and sustainability*, eds. A. H. Alkon, and J. Agyeman, 89–120. Cambridge: The MIT Press.
McCune, N., P. Rosset, T. Cruz, A. Sandíval, and H. Morales. 2016. Mediated territoriality: Rural workers and the efforts to scale out agroecology in Nicaragua. *The Journal of Peasant Studies*. doi: 10.1080/03066150.2016.1233868
Meek, D. 2015a. Learning as territoriality: The political ecology of education in the Brazilian landless workers' movement. *Journal of Peasant Studies* 42 (6): 1179–1200.
Meek, D. 2015b. Towards a political ecology of education: The educational politics of scale in southern Pará, Brazil. *Environmental Education Research*, 21 (3): 447–459.
Meek, D. 2015c. Taking research with its roots: restructuring schools in the Brazilian landless workers' movement upon the principles of a political ecology of education. *Journal of Political Ecology* 22: 410–428.
Meek, D., and L.L. Simonian. 2016. Transforming space and society? The political ecology of education in the Brazilian Landless Workers' Movement's Jornada de Agroecologia. *Environment and Planning D: Society and Space*. doi:10.1177/0263775816667073.
Meek, D., and R. Tarlau. 2015. Critical food systems education and the question of race. *Journal of Agriculture, Food Systems, and Community Development* 5 (4): 131–135.

Meek, D., and R. Tarlau. 2016. Critical food systems education (CFSE): Educating for food sovereignty. *Agroecology and Sustainable Food Systems* 40(3): 237–260.

Molina, M.G. 2013. Agroecology and politics. How to get sustainabillty? About the necessity for a political agroecology. *Agroecology and Sustainable Food Systems* 37: 45–59.

Moll, L. 2002. The concept of educational sovereignty. *Penn GSE Perspectives on Urban Education* 1 (2): 1–11.

Moll, L., and R. Ruiz. 2005. The educational sovereignty of Latino/a students in the United States. In *Latino education: An agenda for community action research*, eds. P. Pedraza, and M. Rivera, 295–320. Mahwah, NJ: Lawrence Erlbaum Associates Publishers.

Morgan, K., and J. Murdoch. 2000. Organic vs. conventional agriculture: Knowledge, power and innovation in the food chain. *Geoforum* 31 (2): 159–173.

Nelson, Alondra. 2011. *Body and soul: The Black Panther Party and the fight against medical discrimination*. Minneapolis: University of Minnesota Press.

Rosset, P.M. 2015. Epistemes rurales y la formación agroecológica en La Campesina. *Revista Ciência & Tecnologia Social* 2 (1): 4–13.

Rosset, P.M., B.M. Sosa, A.M.R. Roque Jaime, and R.A. Ávila Lozano. 2011. The Campesino-to-Campesino agroecology movement of ANAP in Cuba: Social process methodology in the construction of sustainable peasant agriculture and food sovereignty. *Journal of Peasant Studies* 38 (1): 161–191.

Rosset, P., and M.E. Martinez-Torres. 2012. Rural social movements and agroecology: Context, theory and process. *Ecology and Society* 17 (3): 17.

Sawyer, S. 2004. *Crude chronicles: Indigenous politics, multinational oil, and neoliberalism in Ecuador*. Chapel Hill: Duke University Press.

Sbicca, Joshua. 2012. Growing food justice by planting an anti-oppression foundation: Opportunities and obstacles for a budding social movement. *Agriculture and Human Values* 29 (4): 455–466.

Schiavoni, C.M. 2016. Historical, relational and interactive perspectives in food sovereignty research, *ICAS Review Paper Series*, no. 7.

Shepherd, C.J. 2005. Imperial science: The Rockefeller Foundation and agricultural science in Peru, 1940–1960. *Science as Culture* 14 (2): 113–137.

Tarlau, R. 2012. Soviets in the countryside: The MST's remaking of socialist educational practices in Brazil. *Logics of Socialist Education*, 53–72. Dordrecht: Springer.

Tarlau, R. 2015a. How do new critical pedagogies develop? Educational innovation, social change, and Landless Workers in Brazil. *Teachers College Record* 117 (11): 1–36.

Tarlau, R. 2015b. Education of the countryside at a crossroads: Rural social movements and national policy reform in Brazil. *Journal of Peasant Studies* 42 (6): 1157–1177.

Tarrow, S.G. 1994. *Power in movement: Social movements and contentious politics*. Boston: Cambridge University Press.

Thompson, J., and I. Scoones. 1994. Challenging the populist perspective: Rural people's knowledge, agricultural research, and extension practice. *Agriculture and Human Values* 11 (2): 58–76.

White, M.M. 2010. Shouldering responsibility for the delivery of human rights: A case study of the D-Town Farmers of Detroit. *Race/ethnicity: Multidisciplinary Global Contexts* 3(2): 189–211.

White, M.M. 2011. Sisters of the soil: Urban gardening as resistance in Detroit. *Race/ethnicity: Multidisciplinary Global Contexts* 5(1): 13–28.

Wittman, H. 2011. Food sovereignty: A new rights framework for food and nature? *Environment and Society: Advances in Research* 2: 87–105.

Wittman, H. 2015. From protest to policy: The challenges of institutionalizing food sovereignty. *Canadian Food Studies* 2 (2): 174–182.

Wright, E.O. 2012. Transforming capitalism through real utopias. *American Sociological Review* 78 (1): 1–25.

Wright, A., and W. Wolford. 2003. *To inherit the earth: The Landless Movement and the struggle for a new Brazil*. Oakland: Food First!.

Yin, R.K. 2013. *Case study research: Design and methods*. London: Sage.

David Meek Dr. David Meek (PhD University of Georgia, 2014) is an environmental anthropologist, critical geographer, food systems education scholar with an area specialization in Brazil. Dr. Meek theoretically grounds his research in a synthesis of political ecology, critical pedagogy, and agrarian studies. His interests include: sustainable agriculture, social movements, and environmental education. In a series of recent publications, Meek has begun advancing a theoretical framework of the political ecology of education. This perspective illuminates how the reciprocal relations between political economic forces and pedagogical opportunities—from tacit to formal learning—affect the production, dissemination, and contestation of environmental knowledge at various interconnected scales. The various research projects that Professor Meek is involved with provide empirical data to support the advancement of the political ecology of education framework.

Katharine Bradley is a traveling faculty for the International Honors Program: Rethinking Food Security. In collaboration with activists and urban farmers in the Bay Area's food justice movement, she previously helped to start the East Bay Urban Farmer Field School (EBUFFS). Other research topics include participatory research ethics, government support for urban agriculture in the neoliberal era, and race and class dimensions of Community Supported Agriculture. She was a contributor to the Food Dignity Project, which supported her work with the EBUFFS. Food Dignity was supported by Agriculture and Food Research Initiative Competitive Grant no. 2011-68004-30074 from the USDA National Institute of Food and Agriculture.

Bruce Ferguson is a researcher and professor at El Colegio de la Frontera Sur (ECOSUR) in Chiapas, Mexico, where he is a member of the Massification of Agroecology research group. He co-coordinates Laboratorios para la Vida, an action research program that trains educators to use school gardens and local food systems as venues for learning in agroecology, the scientific process, and healthy eating, and for fostering horizontal dialog between local and academic knowledge. His research explores the interface between ecological restoration and agroecology. He teaches graduate courses in restoration ecology and food systems.

Lesli Hoey is an Assistant Professor of Urban and Regional Planning at the University of Michigan (UM), where she is the Coordinator of the Global and Comparative Planning Concentration, Co-Director of the International Planning Case Study Project, and a founding member of the UM Sustainable Food Systems Initiative. She conducts research on food and nutrition planning, policies and programs, which has included research on Bolivia's national Zero Malnutrition Program and the application of collective impact as a tool for promoting policy- and equity-based food systems change in Michigan.

Helda Morales is researcher and professor at El Colegio de la Frontera Sur (ECOSUR) in Chiapas, Mexico, where she is a member of the Massification of Agroecology research group. She co-coordinates, Laboratorios para la Vida, an action research program that trains educators to use school gardens and local food systems as venues for learning in agroecology, the scientific process, and healthy eating, and for fostering horizontal dialog between local and academic knowledge. She also works with urban agriculture from the perspectives of food sovereignty, landscape ecology, and autonomous pest prevention. She teaches graduate courses in food systems and agroecology.

Peter Rosset is professor and researcher at El Colegio de la Frontera Sur (ECOSUR) in Mexico, where he is a member of the Massification of Agroecology research group. He is also a visiting researcher at the Universidade Estadual de Ceará (UECE) and a visiting professor at the Universidade Federal de Ceará (UFC), both in Brazil, and is co-coordinator of the Land Research Action Network (LRAN).

Rebecca Tarlau is a Postdoctoral Scholar in Education at Stanford University. Her ethnographic research agenda has three broad areas of focus: (1) Theories of the state and state-society relations; (2) Social movements, critical pedagogy, and learning; (3) Latin American education and development. Her most recent project examines the educational initiatives of the Brazilian Landless Workers Movement (MST), exploring the movement's attempt to transform public schools across the countryside. Her scholarship engages in debates in the fields of political sociology, international and comparative education, critical pedagogy, global and transnational sociology, and social theory.

Agriculture and Human Values (2019) 36:627–640
https://doi.org/10.1007/s10460-018-9861-9

SYMPOSIUM/SPECIAL ISSUE

Images of work, images of defiance: engaging migrant farm worker voice through community-based arts

J. Adam Perry[1]

Accepted: 23 March 2018 / Published online: 29 March 2018

Abstract

This article addresses a stated need within the food justice movement scholarship to increase the attention paid to the political socialization of hired farm hands in industrial agriculture. In Canada, tackling the problem of farm worker equity has particular social and political contours related to the Canadian horticultural industry's reliance on a state-managed migrant agricultural labour program designed to fill the sector's labour market demands. As Canada's Seasonal Agricultural Worker Program (SAWP) produces relations of 'unfree labour', engaging migrant farm workers in social movement initiatives can be particularly challenging. Critical educational interventions designed to encourage migrant farm workers' contribution to contemporary social movements in Canada must therefore confront the socio-cultural obstacles that constrict migrant farm workers' opportunities to participate as full members of their communities. In this article, I argue that social justice oriented approaches to community-based arts can provide a means for increasing the social movement contributions of farm workers employed through managed labour migration schema such as Canada's SAWP.

Keywords Migrant farm workers · Theatre of the Oppressed · Seasonal Agricultural Worker Program · Guest worker programs · Unfree labour · Canada

Abbreviation

SAWP Seasonal Agricultural Worker Program

Introduction

Community-based cultural production can be a powerful tool for the political socialization of hired farm workers in industrial agriculture. With a focus on Canada's Seasonal Agricultural Worker Program (SAWP), a bi-lateral guest worker regime between Canada, Mexico, and several Caribbean countries, in this article I assess the grassroots organizational potential of directly engaging migrant farm workers in an artistic and collective exploration of their daily lives. In response to a growing literature that laments the lack of attention paid to farm worker equity in the food justice movement, in this article I argue that social justice oriented approaches to community-based arts can provide a way to increase the social movement contributions of farm workers employed through managed migration schema such as Canada's SAWP. I develop this argument from an analysis of participant observation extracted from a community-arts project that I conducted with a small group of SAWP workers in the Southwestern Ontario town of Leamington, a major hub of Canada's horticultural greenhouse industry. This project was grounded in the practice of *Theatre of the Oppressed*, a well-known arts-based adult education tool designed by the late Brazilian educator and theatre director Augusto Boal for engaging non-actors from traditionally marginalized communities as a means of supporting social and political change. While it has been established in the extant literature that engaging SAWP workers in migrant and food justice activism can be an excessively difficult task, it is becoming more widely recognized by food justice activists and scholars that doing so is imperative to developing an inclusive and equitable movement.

While critical artistic interventions have not been widely employed in the Canadian context, these have enjoyed broad popularity in social movement activities with food workers elsewhere, such as in the USA and Brazil. In order to apply these cultural experiments to the Canadian setting, in this article I engage with the conceptual understanding of

✉ J. Adam Perry
contact@adamperry.ca

[1] School of Social Work/School of the Arts, McMaster University, Togo Salmon Hall 410, 1280 Main Street West, Hamilton, ON L8S 4M4, Canada

the SAWP as a guest-worker regime sustained by the incorporation of a state-sanctioned system of 'unfree labour'. Borrowing from the work of feminist philosopher Nancy Fraser, I further argue that the material injustices of unfree labour policies and the cultural narratives that support them profoundly restrict the communicative capacities of SAWP workers, greatly reducing these workers' ability to participate directly in food justice organizing. As a way of addressing the political and cultural 'misrecognition' of Canada's SAWP workers, critical adult educators and food justice advocates should consider alternative, and in particular artistic approaches to grassroots organizing as a way to more effectively incorporate the voices of migrant farm workers in social movement.

The article begins with an examination of farm worker agency in the context of Canada's SAWP. As an entry point, I begin with recent discussions within the food sovereignty scholarship that speak to the need for movement activists and scholars to address the social justice concerns of food labourers. Next, I discuss how the underlying logic of unfree labour relations permeates workers' experiences of daily life in Canada's SAWP. In the following section I examine how the notion of 'parity of participation', a conception of justice developed by feminist critical theorist Nancy Fraser, can help to formulate an approach to organizing SAWP workers that recognizes the need to foster not only the material, but also the cultural resources required to nurture workers' contribution to the food justice movement. I follow this section with an in-depth exploration of the context and results of a theatre-based adult education intervention with a group of SAWP workers employed in Canada's hydroponic greenhouse industry. The article ends with a brief discussion of how arts-based interventions such as the one I describe here can provide an ethical and effective means for workers to contribute to public conversations concerning migrant and food justice.

Farm worker agency in Canada and the logic of unfreedom

In the past decade there has been a growing interest among food justice scholars regarding a perceived disconnect between the food sovereignty movement and social justice concerns related to the exploitation of food labourers (Allen 2008). With notable exceptions (Gottlieb and Joshi 2010; Gray 2014) it is widely recognized that food movement scholarship has not been actively engaged with challenges facing food workers, particularly those who work in agriculture (Levkoe et al. 2016; Minkoff-Zern 2014; Sachs et al. 2014). While recent scholarship has examined the dynamics of unpaid labour on organic farms (Levkoe 2017; Ekers et al. 2016), a wider concern regarding the exploitation of paid farm employees in the context of industrial agriculture remains uncharted (Sachs et al. 2014). This discrepancy has been expressed most succinctly by Minkoff-Zern (2014, p. 86), who claims that "despite the popular attention given to food's provenance, the actual workers who pick and pack our food are consistently overlooked." Similarly, Weiler and colleagues (Weiler et al. 2016) address this tension within the food movement in the Canadian context. Like scholars elsewhere, they claim that hired farm workers are absent from discussions of how to create sustainable food systems. They state that "supporters of these movements have been disengaged, in conflict with, or unsupportive of farmworker equity" (Weiler et al. 2016, p. 2). These various scholars point out how farmworker agency in particular has been either sidelined or not considered, bringing into sharp relief the question, raised by Feagan (2007, p. 29) of "who is in and who is out" when considering the idea of 'local' in food movement activism and knowledge production.

By sidelining the importance of farm worker agency, those involved in the food justice movement are missing an opportunity to understand how workers' interests could potentially shape the negotiation of production politics in agriculture. Concerned scholars call on food researchers to engage in forms of public sociology that support a movement toward social justice for food workers. Allen (2008, p. 160), for example suggests that investigators design "participatory, problem-solving research" that engages real people in their real lives. Similarly Kohl-Arenas et al. (2014, p. 20) make the claim that the enduring inequalities embedded in agricultural work can only be addressed once "the voices and actions of those directly affected are included in the process of designing solutions to the problems they experience." That notwithstanding, for those scholars who have addressed this shortcoming, there is agreement that while it may be imperative to do so, creating spaces for farm workers to participate in political dialogue is "tremendously difficult" (Weiler et al. 2016, p. 12).

The difficulty that Weiler et al. (2016) observe is associated with engaging farm workers in the wider food justice movement has contours that are particular to Canada. While recent scholarship is beginning to connect the struggles of farm workers in Canada to the sustainable food movement (Hjalmarson et al. 2015; Weiler et al. 2016) the issue of agricultural labour in this context has been most widely discussed in the area of migration and labour studies (Basok 2002; Preibisch 2010; Satzewich 1991). The primary focus of this work has been related to social, political, and economic examinations of Canada's Seasonal Agricultural Worker Program (SAWP), a bilateral state-managed 'guest worker' agreement between Canada, Mexico, and several Caribbean countries. A useful conceptual tool that has been employed in the service of interpreting the role that this state managed labour migration schema plays in Canada's horticultural industry is the notion of 'unfree labour' (Choudry and Smith 2016b; Sharma 2012; Thomas 2016). The condition of 'unfree labour' associated with the employment of migrant farm workers in Canada refers to the use of "extraeconomic compulsion... to exact labour power" (Choudry and Smith 2016a, p. 8) from workers, primarily through a restrictive migratory status (Perry 2012b; Satzewich 1991; Sharma 2006).

As a central mechanism of labour control and coercion, migrant farm workers' citizenship restrictions in Canada facilitate various techniques that constrict labour freedoms for workers in both the realms of circulation and production (Lebaron 2015; Preibisch 2010). These include: employer-specific work permits that limit workers' labour mobility; the requirement that workers reside on employer property; the non-authorization of family unification; the requirement that workers return to their home countries at the end of their contracts; and wide-scale restrictions on social and political rights that constrain workers' access to social programs such as employment insurance and English as a Second Language instruction (Basok 2002; Nakache and Kinoshita 2010; Ramsaroop 2016). These policies produce an everyday home/workplace environment in which disciplinary employment practices may thrive, not the least of which are the constant threat of deportation and potential removal from the program as reprisal for resistant or defiant behaviours, including engaging in political activism (Basok et al. 2014; Vosko 2016). While these conditions severely limit the political options available to workers, it is well documented how this state mediated production of unfree labour relations and the incorporation of a trans-migrant labour force have benefited capitalist accumulation in Canadian agriculture since the SAWP's inception in 1966 (Preibisch 2010, 2012). As such, by fashioning workers into "disposable commodities" (De Genova 2005, p. 8) the policies and practices of unfreedom are crucial to production relations in Canada's agri-food sector, making workers' public participation particularly challenging.

Scholars have generally incorporated a political economy approach to examining the production and reproduction of a system of 'unfree labour' in Canada's horticultural industry. Preibisch (2010) for example places an examination of unfree labour in agriculture within a wider context of a liberalized global food economy. She examines how the making of 'unfreedom' is a crucial means by which agricultural businesses can exact greater control over their workers in ways that benefit production. Still other scholars have examined how territorial borders play a key role in "securing and reproducing low wage and 'unfree' labour" (Thomas 2016, p. 24). In this vein, Sharma (2012) examines how nationalist discourses justify the subordination of migrant workers through differential forms of incorporation, thereby containing workers as political subjects while they are simultaneously integrated into the labour market as workers. What these scholars reveal is how extra-economic factors related to restrictive citizenship controls and nationalist discourses are used to extract labour power and pacify workers in the service of capitalist accumulation in Canada's agri-food sector. The result is the production of an agricultural workforce that is more likely to be compliant and less able to resist precarious working and living conditions.

That said, while the structure of unfree labour produces an insidious everyday discipline, there is a need for food movement activists and scholars to recognize that the agency of migrant farm workers is always present, at the very least as an ontological fact (Foucault 1980). Central to the question of worker agency and political subjectivity is how the logic of unfreedom described above influences how workers come to experience everyday life in Canada (Reid-Musson 2017). As Choudry and Smith (2016a, p. 12) observe, the immigration and employment controls that structure the SAWP are "enforced at the level of everyday social life and on the human body." This observation conjures a reading of unfree labour whereby the political and economic structures that produce an unfree condition of work seep into the private and non-work experiences of workers' daily lives (Rogaly 2009; Skrivankova 2010; Strauss 2012). This seepage exposes micro-layers of unfreedom that disturb "workers' control over their own bodies, movement, privacy and social life" (Lebaron 2015, p. 9). The condition of unfreedom to which migrant farm workers in Canada are subject can thus be viewed as a continuum of exploitation that far "exceeds the effects of a political or economic strategy" and profoundly affects workers' private and public lives, their relationships to each other, and the formation of social and cultural identities (De Lissovoy 2015, p. 63).

Research has shown how the logic of unfreedom that underlies workers' social isolation, cultural exclusion, and a lack of access to labour and mobility rights is legitimized through a public discourse that portrays workers as cultural and economic others (Bauder 2008). This is compounded by workers' embodied experiences of racialization in the rural, primarily agricultural, and majority White communities in which they live (Perry 2012b). In other words, underlying the very logic of the program is an order of embodied distribution that classifies SAWP workers as racialized others within what Bakan (2008, p. 5) refers to as "a racialized culture of hegemonic whiteness." Culturally, the logic of unfreedom percolates beneath the surface of workers' relationships, structuring their experiences through a condition of otherness and invisibility, demanding that workers behave like workers. More than just producing a 'reliable' workforce, this underlying logic of subjection communicates the expectation that SAWP workers are "to stay in line and make [themselves] scarce" both inside and outside the workplace (Fanon 2008, p. 94). This produces a hyper-awareness of their surroundings and of their explicit and limited role in Canadian society, thus reifying workers' otherness, shaping their sense of agency, and stifling workers' ability to communicate dissent, both in private and in public.

Recognizing the socio-cultural effects of unfreedom and the necessity to cultivate worker agency are crucial to the development of critical educational interventions. This means developing creative strategies to address the material and cultural obstacles to workers' speaking out about farm worker equity. There is a tremendous need to develop healing and democratic spaces where SAWP workers can experiment with and perform ways of being that belie the underlying logic of unfreedom. Food movement activists can seek out insights and processes from critical pedagogy as a way to harness worker agency and to spark political subjectivities that disrupt the "dominative decorum" that permeates workers' everyday experience of Canada's SAWP (De Lissovoy 2016, p. 346). Currently, there is a growing interest in the application of critical pedagogical methods among food movement activists and scholars. For example, Minkoff-Zern (2014) describes how cultural and educational activities are crucial aspects of developing leadership and solidarity among farm workers themselves.

Leveraging cultural activities to encourage communicative parity

How can cultural and educational interventions as suggested by Minkoff-Zern (2014) address the uneven politics of unfree labour as characterized above? The concept of 'misrecognition' as advanced by feminist political theorist Nancy Fraser (1997, 2013) can assist in developing these connections. Fraser describes misrecognition as "the *material* construction, through the institutionalization of cultural norms, of a class of devalued persons who are impeded from participatory parity" (Fraser 2013, p. 180). For Fraser, misrecognition takes place when social and political institutions regulate a particular group's interactions according to cultural and political norms that "constitute one as comparatively unworthy of respect or esteem" (Fraser 2013, p. 176), thus denying particular categories of people the opportunity to interact with one another as equals. The denial of equitable and unrestrained communication in the public realm to certain groups of people is thus understood as a key element of social injustice. Engaging with Fraser's concept of misrecognition highlights the importance of addressing the problem of voice when attending to farm worker equity in social movement contexts.

In Canada, there are material and ideological factors that impede migrant farm workers from full social and political membership. These contribute to the misrecognition of SAWP workers, and by extension to the scarcity of worker involvement in food justice initiatives. In particular, the disciplinary mechanism of unfree labour as produced through Canada's SAWP ensures that workers are unable to permanently settle in Canada. Workers are instead attached to a single employer throughout the duration of their contracts. These material restrictions limit workers' labour and geographic mobility, increase workers' risks of experiencing labour standards abuses, and also limit workers' ability to actively contribute to formal and grassroots debates that may potentially influence public policy. The cultural narratives that render SAWP workers as outsiders support these institutional relations, thus legitimating the permanent alienation of SAWP workers from having their voices heard in social and political forums. The dynamics of unfree labour, in tandem with the ideological discourses that legitimize them, thus produce a "vicious circle of cultural and economic subordination" (Fraser 1997, p. 15) that codes the patterns shaping the range of communicative possibilities available to SAWP workers in their everyday relationships. The issue of paid farm hands' absence from Canada's food movement is thus not just a problem of advocates' lack of access to workers, but is also a consequence of the deeper problem of workers' material and cultural exclusion from social and political life.

In response to the problem of misrecognition, Fraser posits a conception of justice that is grounded in the principle of "parity of participation", which "requires social arrangements that permit all (adult) members of society to interact with one another *as peers*" (Fraser 2013, p. 164). In order to increase the contribution of farm workers in social movement, migrant and food justice initiatives would do well to incorporate activities that encourage the type of communicative parity imagined by Fraser. This means developing educational environments that address the underlying cultural values that serve to justify the broad denial of workers from the rights and protections of full citizenship. In keeping with Fraser's notion of parity of participation, this must include creating spaces where migrant farm workers can come together to challenge the injuries suffered as a consequence of their pervasive exclusion from public life. Canada's food movement must develop creative opportunities for workers to contest cultural values that reify migrant farm worker subordination as the norm.

Cultural and creative activities are not unusual in social movement organizing. On the contrary, interweaving community organizing with artistic practices, such as collective storytelling (Newman 2006) and performance (Hsia 2010), is increasingly common as a potentially effective grassroots approach to strengthening the communicative potential of people from traditionally marginalized groups. Within the field of critical adult education, the arts have been proposed as an important vehicle to cultivate and communicate counter-hegemonic ideas and to develop relationships of solidarity and connection in social movement contexts (Butterwick and Roy 2016; Brookfield and Holst 2010; Choudry 2015). Notably, migrant and food justice movements alike have mobilized artistic forms of "cultural organizing" to claim public space and to honour both farm workers and transnational migrants in various international settings as "creative and purposeful human beings" (Kohl-Arenas et al. 2014, p. 12). Both the Landless Farm Worker Movement in Brazil and Florida's Coalition for Immokalee Workers for example have incorporated storytelling, music, and theatricality as a way to build solidarity and forge common identity among agricultural workers and the general public (Gouge 2015; Hammond 2014; Issa 2007). Other examples, such as those that have incorporated artistic exploration with transnational marriage migrants in Taiwan (Hsia 2010; Liang 2016) and Filipina live-in caregivers in Canada (Butterwick et al. 2015), have integrated performance as a way to create a space where communal migrant subjectivities may be cultivated. As seen through the lens of Nancy Fraser's conceptualization of justice, these examples highlight how collective artistic creation can leverage often-untapped cultural resources to address the problem of misrecognition and to support parity of participation among traditionally marginalized communities. How can cultural activities be mobilized with hired farm hands who are affected by unfree labour policies, such as SAWP workers in the context of Canada's horticultural industry? In what follows I discuss the trajectory and results of my own efforts to carve out a space where SAWP workers are encouraged to contribute their voice to migrant and food justice activism through cultural and artistic expression.

Engaging migrant farm workers in community-based arts

The reflection offered in this article is grounded in my 15 years of experience as an adult educator, advocate, and later, as a university-based researcher working with SAWP workers. My work with migrant workers began with my position as a labourer-teacher with Frontier College, a well-established literacy organization in Canada. In this position I worked as both a farmworker and as a volunteer English as a Second Language and literacy instructor. In order to sustain my volunteer work, for 3 years I worked and lived alongside migrant farm workers on tobacco and fruit farms in the agriculture-rich region of Southwestern Ontario. My proximity to workers as a fellow co-worker greatly informed my work as an advocate and as an educator (Perry 2013). During these years I spent my evenings and weekends delivering student-centred language lessons to interested co-workers and their friends. Lessons were focused on fulfilling workers' daily communication needs in the local language, providing a practical foundation for encouraging worker independence.

After 2 years as a language instructor, I was approached by a group of advanced language learners in the municipality of Leamington (pop. 27,595), a crucial hub for Canada's greenhouse industry. In 2015 (the most recent numbers available) Leamington was host to 5743 migrant workers, the vast majority employed in the agri-food industry (Government of Canada 2018). This group of workers wanted my help to write and perform a collaboratively created play about their lives as migrant farm workers in Canada. This group made it clear that they wanted to perform in English, and that their intended audience would be local Canadians. Feeling socially and culturally isolated in the town where they lived, the group's goal was to employ a creative outlet to educate locals about their daily lives. Over the course of a summer, this group of workers met twice a week after work, often commuting long distances to town by bicycle, to create and rehearse a short play that was eventually performed for a mixed audience of SAWP workers and local Canadians at a community centre. While this project provided a means for workers to express themselves creatively, workers' decision to target an audience that included non-migrant workers introduced a pedagogical limitation. Namely, with a Canadian audience in mind, this group decided to develop and publically share a story about life as a migrant worker that purposefully did not challenge the accepted status quo, so as to avoid the known social risks associated with workers publicly expressing themselves as political subjects. With the worry that SAWP employers may be present in the audience (and some were), the group chose not to document workers' real-life struggles and instead crafted a light-hearted love story, thus portraying the SAWP as benign and workers themselves as gentle and fun-loving neighbours.

The experience of leading this project was both invigorating and frustrating. On the one hand, having the chance to help workers to shape and share their own stories in a public forum was energizing. On the other hand, being constrained with regards to the types of stories we could share as a group felt like a missed opportunity. While the production of this play was considered an important and distinctive cultural performance that communed a rich emotional depiction of life in Canada's SAWP (Hill 2005), it did not embrace an explicit analysis of how relations of dominance imbue workers' experiences of work and migration. Could a cultural activity such as the one proposed by this group of SAWP workers not be a potentially effective approach to public and political pedagogy, and contribute in some way to established migrant and food justice movements? To the contrary, this project was illustrative of the challenges associated with the cultural and material obstacles to inviting SAWP workers to contribute their voices to political struggle. While artistic activities associated with critical adult education may offer a means of encouraging workers to express a collective subjectivity (Brookfield and Holst 2010; Hsia 2010; Newman 2006), in this case the risks associated with public displays of communal agency suppressed even the perception that workers were reflecting on and critiquing dominant social relations. Several years later, as a university-based researcher, I decided to try this play-making experiment a second time with a different group of workers. With an eye to liberating workers from the gaze of employers and the wider Canadian population, this second project was designed from the outset to be by and for workers themselves. Given the results of the previous play-making experiment, I felt that workers interested in cultural production needed a private worker-only space to experiment—a place that was free from outside scrutiny (see Flores and Garcia 2009). The idea was to create a space where workers could bear communal witness to life in the SAWP and where workers' experiential knowledge of unfree labour could be nurtured. A crucial question animating this decision was related to how an SAWP worker-only arts-collective could provoke a grassroots cultural interchange that could address the problem of misrecognition as it is expressed through Canada's managed labour migration system.

A note on methods

I base the analysis presented in this article on a series of theatre workshops conducted with a group of SAWP workers from Mexico working in Canada's hydroponic greenhouse industry. Methodologically, this project was committed to both emphasizing the agency of participants and to producing social science research that does not simply document the social world, but which is active in its very production (Denzin 2003; Law and Urry 2004). On both of these fronts, a performative approach to qualitative inquiry offered a rich and provocative invitation toward engaging with farm workers. To better understand how the application of arts-based workshops can be harnessed in food movement activism, this article focuses on cooperative meaning making and solidarity learning as these occurred during the workshops themselves and the performance that followed.

The theatre-creation workshop sessions were developed from well-established community-based approaches to collective theatre creation, primarily the *Theatre of the Oppressed*, a social justice-oriented theatrical technique developed specifically for use with non-actors by the late Brazilian theatre director and educator Augusto Boal (1985, 2002). Among activists and arts educators, Boal's *Theatre of the Oppressed* is often portrayed as an artistic technique for critically examining hegemonic social relations with and among traditionally marginalized groups (Barak 2016; Hsia 2010). As per the process of Boal's work, our sessions involved the group coming together to examine their everyday lives through physical movement exercises, theatrical improvisations, in-depth discussions, storytelling, collective play creation, and community-based performance. The project incorporated three distinct phases: embodied inquiry, play-creation, and public performance (Perry 2018). The first phase was grounded in Boal's technique of 'image theatre' (Boal 2002, pp. 139–147). In this phase workers created static body images of their lives in the SAWP. By working in mostly silence and focusing primarily on workers' bodies, this phase was designed to emphasize workers' non-verbal resources, including emotions, memories, and workers' perceptions of embodied interactions with others. Intended to spark collective memories and new interpretations of life in the SAWP, these images became a catalyst for group discussions in which workers shared and compared their own personal life histories (Perry 2012a). This phase was followed by the creation of a short play. The collectively created play was based on the images that were generated in the previous phase and the discussions that followed. Using a patchwork of methods derived from Boal's toolbox, such as collective improvisation and character development, participants constructed a series of dramatic scenes that examined the challenges associated with a worker's first year in the program.

An integral aspect of this project was to provide adequate space for workers themselves to collectively improvise their own representations of daily life as migrant farm workers. In the interests of respecting workers' own interpretations of life in Canada's SAWP, participants were therefore fully engaged in all aspects of developing artistic depictions (Sinding et al. 2008). This meant that all participants were involved in artistic creation and that all of the images, characters, dialogue, and scenes that workers created were examined in relation to each individual worker's lived experiences, worldviews, and personal beliefs. All of the voices present were therefore urged to partake in a conversation of how to best portray the intimate realities of life for participants. In practical terms, this meant encouraging participants to pause the workshop activities at any time in order to discuss in more detail the topic at hand. The process of engaging with a multiplicity of worker voices deepened the ethical grounding of the project and strengthened the accuracy of artistic representations prior to public scrutiny (Conquergood 1985; Mienczakowski 1999). Unlike the previous play that was created for a mixed audience of workers and Canadians, the play that participants created expressly probed some difficult material, such as uneven power relations in the workplace, co-worker harassment, problematic interactions with program officials, and loneliness. After the play was created and rehearsed, we invited an audience of 25 fellow workers to attend the final performance. The presentation was designed as a community forum. This involved incorporating a facilitated discussion meant to elicit the audiences' responses to both the content covered in the play as well as the experience of being involved in a worker-led dialogue on their lives in the SAWP.

In an effort to access workers most obviously denied the status as 'peer' as per Fraser's conceptualization of misrecognition, my goal was to enlist workers who were not already involved in community organizing. The rationale was to encourage workers who did not already have the experience of expressing themselves in a public and overtly political way to experiment with the development of a communal voice. While I was no longer volunteering as a language instructor at the time, I was given permission by Frontier College to attend several of their Leamington classes in order to pitch my idea to potential participants. In these presentations I told workers about my own experiences as a farmworker and volunteer advocate, and told them about the previous theatre-making project. I informed workers that the point of the current project was to bring workers together to build community and to create and share stories amongst themselves that were relevant to their true-to-life struggles as migrant farm workers in Canada. I explained that the project would be limited to workers only, and that no members of the local community would be involved in either the workshops or the public performance. In this way, I was able to offer workers an assurance of relative confidentiality.

Interested workers were also made aware that this endeavour was associated with a university-based research project that examined the efficacy of artistic interventions to create meaningful stories that support worker solidarity and parity of participation. On this front, I asked workers' permission to observe their interactions during the theatre-creation workshops and to include their anonymized exchanges in future reports and publications. Audience members (who were also SAWP workers) were similarly informed of the research dimension of the project, and were likewise asked for their consent to participate. In the end, I enlisted a group of nine SAWP workers from Mexico to contribute to a theatre creation and performance project that lasted throughout a complete agricultural season. The workers who were most eager to take part in this project were individuals who had expressed a desire to develop social connections outside of work through involvement in activities separate from their working lives. For these amateur worker-artists, our weekly workshops were both an opportunity to socialize with like-minded workers and a means to express parts of their identities that normally lie dormant throughout their sojourns in Canada. Those workers who stated their motivation for participating in the project discussed how they found meaning in creating art that mirrored their own experiences and those of their fellow SAWP workers. One worker (Joaquín)[1] expressed this sentiment thus:

> The project is important because the play expresses the situations that we are living as migrant workers.[2]

For these workers having the chance to express themselves artistically was especially appealing. In the words of one worker (Enrique):

> I liked the idea of being in a play about Mexicans. I always wanted to participate in something like this. Well, when you invited me it got me really excited.

Drawing from my own observations of participant interactions during this project, in what follows I offer a glimpse of how artistic educational interventions with migrant farm workers may be one way to incorporate the voices of food labourers into migrant and food justice movements. By focusing most of the analysis on the first, 'image theatre' phase of this project, I examine how simple theatrical explorations can offer a unique means of accessing workers' embodied experiences of work, unfree labour, and social struggle. I reveal how a thoughtful exploration of workers' experiential knowledge of life in Canada's SAWP can facilitate critical collective discussions of how power infiltrates workers' everyday lives. The discussion below demonstrates how in-depth investigations of the memories and emotions that undergird workers' experiences of manual agricultural labour produced an opportunity to augment participants' capacity to communicate a collectively produced reflection on worker camaraderie, worker agency and the logic of unfree labour.

Images of work, images of defiance

Work

> You have to live what we live in your own body to know what we feel. The play more than anything provides a reflection on what it is we are living through here in Canada as workers... on what we are *really* living through.

An (anonymous) SAWP worker expressed this provocative statement during the post-performance discussion at the end of the arts project described in this article. This quote demonstrates the importance of engaging with farm workers' experiential knowledge of life in Canada's SAWP as a way

[1] I provided workshop participants with pseudonyms.

[2] Worker quotes have been translated to English from the original Spanish.

of learning about the lived realities of unfree labour relations. For critical adult educators, finding a way to encourage workers to physically interrogate the social relations that shape their daily lives is crucial to incorporating the voices of SAWP workers in food movement activism, as is suggested in the literature reviewed above. In keeping with this goal, the first phase of this project involved facilitating 'image theatre' exercises with the group. Practitioners of the *Theatre of the Oppressed* describe the technique of image theatre as "a series of wordless exercises in which participants create embodiments of their feelings and experiences" (Schutzman and Cohen-Cruz 1994, p. 237). Given this technique's emphasis on accessing memories, emotions and other physical experiences that may be difficult to articulate clearly in spoken language, I chose image theatre specifically because of its potential to engage workers' tacit and embodied knowledge of everyday life (Perry 2018).

During the first few sessions, an explicit focus on embodied knowledge did not immediately produce in-depth explorations of what could be characterized as workers' experiential knowledge of unfree labour relations. Rather, when workers were offered the chance to use their bodies to communicate information about their lives in the SAWP, they instinctively created images that depicted the simple everyday actions and situations associated with physical agricultural labour. For example, during the early stages of the project, workers created images of picking tomatoes, tying the vines of tomato plants, and of manipulating complex greenhouse machinery. This was not surprising, seeing as though SAWP labourers often work upwards of 60-h weeks and are drawn from the ranks of racialized global labouring classes (Basok 2002; Smith 2016). These are workers for whom rich and varied experiences of manual labour are deeply embedded in their embodied understandings of daily life. SAWP workers' day-to-day activities at work were therefore easily recreated through theatrical image making. In the words of one project participant (Mauricio):

> We do these activities all day everyday. These actions are impossible to forget.

These images offered more than just a surface-level representation of participants' working lives, however. By coming together on a weekly basis to engage in an artistic exploration workers developed affective camaraderie grounded in their own intimate experiences of agricultural work, which provided an effective means of developing a communal subjectivity. For example, during the workshops participants discussed openly how creating theatrical images of work sparked individual memories of their lives in the program that merged with the collective memory of the group. This was particularly true for one worker (Enrique), who was in his first year in the program at the time of research and was particularly struck by how the experiences of older workers mirrored his own. Reflecting on an image created by one of his more senior co-workers that depicted his employer yelling at him on his first day of work, Enrique commented:

> I remembered when I first came into the program. Joaquín's story was almost exactly the same as what happened to me.

Connecting the individual memory of work to the collective memory of exploitation may thus have supported the development of a communal understanding of the SAWP experience. The workshops thereby provided a space for workers to develop a shared sense of responsibility and care, thus triggering an important interruption of workers' everyday experiences of social exclusion and isolation. For example, during these initial theatre workshops it was not unusual for older workers to stop the action in order to provide council to younger workers on how to develop practical strategies to protect their bodies from long-term injuries. This collective interrogation of how life in the SAWP could be represented through a commonly understood physical vernacular provided a central foundation for surfacing workers' embodied knowledge of broader social relations.

Defiance

In an effort to deepen the analysis, these embodied images of work were "brought to life" through interactive exercises known to *Theatre of the Oppressed* practitioners as 'dynamizing' activities (Schutzman and Cohen-Cruz 1994, p. 237). With varying degrees of success, and using assorted improvisational tools, I encouraged participants to reflect on the internal and external forces that animated the collectively created physical images of agricultural labour. For example, during this phase participants improvised the development of different characters, such as SAWP employers, government officials, spouses, and fellow workers. These improvisations presented a window to interrogating the underlying emotions, motivations, and political interests that determine workers' everyday experiences of work. In one activity that proved particularly fruitful, in the process of demonstrating previously created images of greenhouse work, I asked participants to express their character's 'inner monologues' (Boal 2002, p. 207). In this exercise, participants voiced the inner thoughts of the characters (i.e. workers, employers, consular officials) involved in the images of work that they had already created. These otherwise innocuous images of agricultural labour were thus transformed. From these images participants created polyphonic soundscapes that evoked a potent account of how power, as filtered through the policies and practices of unfree labour, shapes the embodied everyday experience of life in Canada's SAWP. As an example, the following internal monologues developed from an image of greenhouse work are illustrative.

One participant, who was playing the part of an employer, voiced his complaints about his Mexican employees working too slowly and decided that he would threaten to send them back to Mexico in order to make them work faster. Another participant, who was playing a worker in his first SAWP season, extracted a litany of profane language thrown in the direction of his employer, who he felt was treating him unfairly. The next participant, playing the part of a worker, spoke to his spouse about how much he missed her, and that he hoped working in Canada would provide financial opportunities for their family. A fourth participant, also playing the role of a worker, expressed his fear that the employer would punish him and his fellow co-workers because one member of his crew was working too slowly. Voicing the private concerns of fictional (but true-to-life) SAWP characters that they themselves created encouraged participants to critically examine how the policies and practices of unfree labour may infiltrate workers' bodies and structure their everyday relationships.

Unlike earlier discussions that focused primarily on building workers' long-term physical resilience to the bodily hardships associated with farm work, these moments provoked conversations in which workers shared and compared memories of unjust treatment and debated the risks associated with speaking out. These discussions in turn provided the groundwork for developing embodied images of defiance that could then be incorporated into migrant and food justice organizing. To illustrate: at the time this phase of the project was taking place, the volunteer-led migrant justice collective, *Justicia for Migrant Workers* was organizing a campaign called the 'Pilgrimage to Freedom', a summer-long action which involved grassroots activists and migrant farm workers staging events across the province of Ontario. Event organizers invited interested workers from our group to join a small group of worker-activists to create dramatic images that could be incorporated into the caravan's communications strategy, such as campaign posters and banners. This invitation provided a means for this group of artist-workers to contribute to public and political pedagogy while at the same time maintaining their anonymity. Prior to this gathering, I invited workers to bring objects that for them represented different aspects of their lives in Canada. On the day of the workshop we had a motley collection of farm-related objects, including various small agricultural tools as well as an assortment of tomatoes and cucumbers. Using these objects as props, throughout the day, this blended group of workers collectively created images meant to represent both the realities of workers' lives and the broader struggle for farm worker equity.

By the end of this session, the collective of workers had settled on one particular image that we felt captured workers' experiential knowledge of manual work, unfree labour, and social struggle. In the image (Fig. 1), a hand holding a tomato represents manual work, a chain wrapped around the worker's hand and arm symbolizes unfree labour, and a fist raised in defiance embodies the idea of social struggle. This image was transformed into a digital image that was later attached to *Justicia for Migrant Workers's* 2016 'Harvesting Freedom' campaign (Justicia for Migrant Workers 2017), a Canada-wide migrant justice event that commemorated the 50-year anniversary of Canada's SAWP (Fig. 2) (Justicia for Migrant Workers 2017). Depictions of manual labour were thus transformed into embodied images that contributed to on-the-ground migrant and food justice activism. This example demonstrates how a recognized arts-based educational practice can be taken up in the service of increasing the communicative capacity of migrant farm workers in the context of community organizing. The use of Augusto Boal's technique of image theatre was particularly useful in accessing workers' embodied knowledge of work, unfree labour, and social struggle. This technique, which focused on engaging workers' bodies in a collective investigation of life in Canada's SAWP, provided a creative means to spark not only a sense of affective camaraderie and communal subjectivity amongst workers not normally engaged in social movement, but also to provoke conversations that may otherwise not have taken place. Crucial to this approach is how arts workshops provided a grassroots mechanism for cooperatively examining some of the most salient and richest material available for engaging workers in struggles for justice, namely workers' own direct experiences of physical labour.

A note on public performance

The eventual performance derived from the images described here provides further evidence as to the potential and the obstacles for community-based artistic practice to contribute to discussions of farm worker equity. This performance took place near the end of the agricultural season. Fellow SAWP workers who were in attendance were invited to participate in a discussion on the utility of artistic creation to communicate workers' impressions of life in the program. These workers' comments highlight how artistic expression can provoke strong feelings regarding workers' own perceptions of whether or not they are capable of or even entitled to expressing discontent. On the one hand, some workers expressed consternation on witnessing workers express oppositional subjectivities on the stage. The following anonymous audience comments reflect this position:

> If we present this to the public, well, I don't know. Like what do you think? What would the Canadian government say? What would the Mexican consulate say? They might just grab you and ask 'who gave you the permission to have this idea?'

Fig. 1 Image of defiance

> We have a situation here, because Canadians will say, 'who let them do this'? The bosses will say, the Mexican consulate will say, the Mexican Ministry of Labour will say, 'who let them do this?' The bosses would not like to see this. They would send us all back to Mexico.

On the other hand, some workers felt as though play creation and performance could provide an effective way to increase workers' communicative capacity. The following comments reflect this position:

> The play releases the burden that we all have, to let it go, to express it. It is a way to rebel against the situation that we are all living, because the other option is to just remain silent.

> Do you have courage, or what? I have courage, and for me, I would perform this play in public because everything is shown—the good *and* the bad. Think about the consequences that this play could have. If people could start to see all these things, I don't know how far it would go, but it would create a real commotion.

Workers' performative subversion of the accepted norms of what it means to be a migrant farm worker in Canada through theatre creation sparked an emotional dialogue among those workers present in the audience. That audience members were so animated in their responses to the artistic expression of their fellow workers demonstrates how deeply enmeshed this project was in bringing to light the politics of worker representation in social struggle.

Fig. 2 2016 harvesting freedom campaign logo

While no clear resolution came out of this discussion, its very occurrence represents an incremental step toward participatory parity in the struggle to incorporate migrant farm worker voice into food justice organizing. Workers' collective exploration of manual agricultural work, as expressed through the artistic engagement of workers' actual bodies, presented an opportunity to aesthetically tap into workers' emotions, memories, and embodied experiences of unfree labour. This initiative offered a group of farm workers the time and space to reflect on their lives, and a relatively low-risk platform from which they could explore how they may want their voices heard within the contemporary food and migrant justice movements. Borrowing insights from Fraser (1997, 2013), this project provided a cultural platform for migrant farm workers to interact as peers in such a way that invokes the prospective of engaging with participatory parity in the migrant justice and food sovereignty movements.

Conclusion

Are there aspects of this small-scale community-based theatre project that address the problem, as articulated in the literature, of connecting the struggles of actual farm workers to the wider food justice movement? In the context of Canadian agri-business, this question must be framed in relation to the dynamics of unfree labour, and how these shape migrant farm workers' experiences of life in Canada's Seasonal Agricultural Worker Program. The notion of unfree labour that I examine in this article is gaining currency of late in the labour migration scholarship coming out of the Canadian context, and for good reason. Engaging with the idea of unfree labour provides a useful conceptual tool for understanding how restrictive citizenship controls and nationalist discourses are put to work in extracting labour control and pacifying workers in Canada's agricultural sector. In addition to restricting the political, geographic and labour mobility of SAWP workers, the logic of unfreedom that undergirds the structure of transnational agricultural employment in Canada is an insidious form of everyday discipline. The politics of unfree labour that variously constrain SAWP workers' opportunities to participate as full community members through mechanisms such as the constant threat of deportation, are thus sutured to workers' embodied experiences of daily life. I thereby introduce the concept of misrecognition as posited by Nancy Fraser as a way to further deepen our understanding of how the cultural and economic politics of unfree labour infiltrate SAWP workers' everyday relationships. By inspecting Canada's SAWP through the lens of misrecognition, I make the case that the politics of unfree labour produce communicative constraints for farm workers that profoundly affect their opportunities to interact as equals in Canadian society. This presents a grave problem for Canadian food justice advocates who are interested in engaging farm workers in public conversations around farmworker equity and food sovereignty. However, in this article I argue that Canada's migrant and food justice advocates must address the lack of communicative resources available to SAWP workers in order to address problems related to the exploitation of food labourers. Educational interventions designed to encourage migrant farm workers' contribution to social movements must therefore tackle the socio-cultural obstacles that constrict migrant farm workers' opportunities to participate as full members of their communities. When designed thoughtfully, social justice oriented approaches to community-based arts can provide one way of accomplishing this objective.

Acknowledgements The author wishes to gratefully acknowledge the helpful comments of the three anonymous reviewers during the preparation of this manuscript. The funding was supported by Social Sciences and Humanities Research Council of Canada (Grant No. 752-2011-2626).

Compliance with ethical standards

Ethical approval All procedures performed in studies involving human participants were in accordance with the ethical standards of the University of Toronto and with the 1964 Helsinki declaration and its later amendments or comparable ethical standards.

Informed consent Informed consent was obtained from all individual participants included in the study.

References

Allen, P. 2008. Mining for justice in the food system: Perceptions, practices, and possibilities. *Agriculture and Human Values* 25(2): 157–161.

Bakan, A. B. 2008. Reconsidering the underground railroad: Slavery and racialization in the making of the Canadian state. *Socialist Studies* 4(1): 3–29.

Barak, A. 2016. Critical consciousness in critical social work: Learning from the Theatre of the Oppressed. *British Journal of Social Work* 6(1): 1776–1792.

Basok, T. 2002. *Tortillas and tomatoes: transmigrant Mexican harvesters in Canada*. Montreal: McGill-Queen's University Press.

Basok, T., D. Bélanger, and E. Rivas. 2014. Reproducing deportability: Migrant agricultural workers in South-western Ontario. *Journal of Ethnic and Migration Studies* 40(9): 1394–1413.

Bauder, H. 2008. Foreign farm workers in Ontario (Canada): Exclusionary discourse in the newsprint media. *Journal of Peasant Studies* 35(1): 100–118.

Boal, A. 1985. *Theatre of the oppressed*. New York: Theatre Communications Group.

Boal, A. 2002. *Games for actors and non-actors*. London: Routledge.

Brookfield, S., and J. Holst. 2010. *Radicalizing learning: Adult education for a just world*. San Francisco: Jossey-Bass.

Butterwick, S., and C. Roy. 2016. *Working the margins of community-based adult learning*. New York: Springer.

Butterwick, S., M. Carrillo, and K. Villagente. 2015. Women's fashion shows as feminist transformation. *The Canadian Journal for the Study of Adult Education* 27(2): 79–99.

Choudry, A. 2015. *Learning activism: The intellectual life of contemporary social movements*. Toronto: University of Toronto Press.

Choudry, A., and A. Smith. 2016a. Introduction: Struggling against unfree labour. In *Unfree labour? Struggles of migrant and immigrant workers in Canada*, eds. A. Choudry, and A. Smith, 1–20. Oakland: PM Press.

Choudry, A., and A. Smith, eds. 2016b. *Unfree labour? Struggles of migrant and immigrant workers in Canada*. Oakland: PM Press.

Conquergood, D. 1985. Performance as moral act: Ethical dimensions of ethnography of performance. *Literature in Performance* 5(2): 1–13.

De Genova, N. 2005. *Working the boundaries: Race, space, and "illegality" in Mexican Chicago*. Durham: Duke University Press.

De Lissovoy, N. 2015. *Education and emancipation in the neoliberal era: Being, teaching, and power*. New York: Palgrave Macmillan.

De Lissovoy, N. 2016. Race, reason and reasonableness: Toward an "unreasonable" pedagogy. *Educational Studies* 52(4): 346–362.

Denzin, N. K. 2003. *Performance ethnography: Critical pedagogy and the politics of culture*. Thousand Oaks: Sage Publications.

Ekers, M., C. Z. Levkoe, S. Walker, and B. Dale. 2016. Will work for food: Agricultural interns, apprentices, volunteers, and the agrarian question. *Agriculture and Human Values* 33(3): 705–720.

Fanon, F. 2008. *[1952] Black skin, white masks*. New York: Grove Press.

Feagan, R. 2007. The place of food: Mapping out the "local" in local food systems. *Progress in Human Geography* 31(1): 23–42.

Flores, J., and S. Garcia. 2009. Latina testimonios: A reflexive, critical analysis of 'Latina space' at a predominantly White campus. *Race, Ethnicity and Education* 12(2): 155–172.

Foucault, M. 1980. *Power/knowledge: Selected interviews and other writings 1972–1977*. New York: Pantheon Books.

Fraser, N. 1997. *Justice interruptus: Critical reflections on the 'post-socialist' condition*. New York: Routledge.

Fraser, N. 2013. *Fortunes of feminism: From state-managed capitalism to neoliberal crisis*. London: Verso.

Gottlieb, R., and A. Joshi. 2010. *Food justice*. Cambridge: MIT Press.

Gouge, M. C. 2015. Human rights in play, transnational solidarity at work: Creative playfulness and subversive storytelling among the Coalition of Immokalee Workers. *Critical Sociology* 42(6): 861–875.

Government of Canada. 2018. Annual labour market impact assessment statistics 2008–2015 by urban area. https://www.canada.ca/en/employment-social-development/services/foreign-workers/reports/2014/lmia-annual-statistics/urban-area.html. Accessed 10 March 2018.

Gray, M. 2014. *Labor and the locavore: The making of a comprehensive food ethic*. Berkeley: University of California Press.

Hammond, J. L. 2014. Mística, meaning and popular education in the Brazilian Landless Workers Movement. *Interface: A Journal for and about Social Movements* 6(1): 372–391.

Hill, S. 2005. Migrant workers take to stage. *Windsor Star*, August 19.

Hjalmarson, E., R. Bunn, A. Cohen, E. Terbasket, and L. Gahman. 2015. Race, food, and borders: Situating migrant struggle in the Okanagan Valley, British Columbia. *Journal of Agriculture, Food Systems, and Community Development* 5(4): 77–82.

Hsia, Hsiao-Chuan. 2010. The subjectivation of marriage migrants in Taiwan: The insiders' perspectives. In *Learning from the ground up: Global perspectives on social movements and knowledge production*, eds. A. Choudry, and D. Kapoor, 101–120. New York: Palgrave Macmillan.

Issa, D. 2007. Praxis of empowerment: Mística and mobilization in Brazil's landless rural workers' movement. *Latin American Perspectives* 34(2): 124–138.

Justicia for Migrant Workers. 2017. Harvesting freedom. https://harvestingfreedom.org/. Accessed 24 Oct 2017.

Kohl-Arenas, E., M. M. Nateras, and J. Taylor. 2014. Cultural organizing as critical praxis: Tamejavi builds immigrant voice, belonging, and power. *Journal of Poverty* 18(1): 5–24.

Law, J., and J. Urry. 2004. Enacting the social. *Economy and Society* 33(3): 390–410.

Lebaron, G. 2015. Unfree labour beyond binaries: Insecurity, social hierarcy and labour market restructuring. *International Feminist Journal of Politics* 17(1): 1–19.

Levkoe. 2017. Engaging the tensions of ecological internships: Considerations for agroecology and sustainable food systems movements. *Agroecology and Sustainable Food Systems*. https://doi.org/10.1080/21683565.2017.1347120.

Levkoe, C. Z., N. McClintock, L.-A. Minkoff-Zern, A. K. Coplen, J. Gaddis, J. Lo, and A. M. Weiler. 2016. Forging links between food chain labor activists and academics. *Journal of Agriculture, Food Systems, and Community Development* 6(2): 129–142.

Liang, P. 2016. Interweaving ethics and aesthetics: Marriage migrants and theatre making in the Asian transnational space. *Inter-Asia Cultural Studies* 17(3): 334–356.

Mienczakowski, J. 1999. Ethnography in the hands of participants: Tools for dramatic discovery. In *Explorations of methodology*, eds. G. Walford, and A. Massey, 145–161. Bingley: Emerald Group Publishing Limited.

Minkoff-Zern, L.-A. 2014. Challenging the agrarian imaginary: Farmworker-led food movements and the potential for farm labor justice. *Human Geography* 7(1): 85–101.

Nakache, D., and P. J. Kinoshita. 2010. *The Canadian temporary foreign worker program: Do short term economic needs prevail over human rights concerns?* Montreal: Institute for Research on Public Policy.

Newman, M. 2006. *Teaching defiance: Stories and strategies for adult educators*. San Francisco: Jossey-Bass.

Perry, J. A. 2012a. A silent revolution: 'Image theatre' as a system of decolonization. *Research in Drama Education: The Journal of Applied Theatre and Performance* 17(1): 103–119.

Perry, J. A. 2012b. Barely legal: Racism and migrant farm labour in the context of Canadian multiculturalism. *Citizenship Studies* 16(2): 189–201.

Perry, J. A. 2013. Living and learning through solidarity and struggle: Assessing the informal learning of Frontier College labourer-teachers. In *Volunteer work, informal learning and social action*, eds. F. Duguid, K. Mündel, and D. Schugurensky, 79–99. Rotterdam: Sense Publishers.

Perry, J. A. 2018. Play-making with migrant farm workers in Ontario, Canada: A kinesthetic and embodied approach to qualitative research. *Qualitative Research*. https://doi.org/10.1177/1468794117743463.

Preibisch, K. 2010. Pick-your-own labor: Migrant workers and flexibility in Canadian agriculture. *International Migration Review* 44(2): 404–441.

Preibisch, K. 2012. Migrant workers and changing work-place regimes in contemporary agricultural production in Canada. *International Journal of Sociology of Agriculture and Food* 19(1): 62–82.

Ramsaroop, C. 2016. The case for unemployment insurance benefits for migrant agricultural workers in Canada. In *Unfree labour? Struggles of migrant and immigrant workers in Canada*, eds. A. Choudry, and A. Smith, 105–122. Oakland: PM Press.

Reid-Musson, E. 2017. Intersectional rhythmanalysis: Power, rhythm, and everyday life. *Progress in Human Geography*. https://doi.org/10.1177/0309132517725069.

Rogaly, B. 2009. Spaces of work and everyday life: Labour geographies and the agency of unorganized temporary migrant workers. *Geography Compass* 3: 1975–1987.

Sachs, C., P. Allen, A. R. Terman, J. Hayden, and C. Hatcher. 2014. Front and back of the house: Socio-spatial inequalities in food work. *Agriculture and Human Values* 31(3): 3–17.

Satzewich, V. 1991. *Racism and the incorporation of foreign labour: Farm labour migration to Canada since 1945*. New York: Routledge.

Schutzman, M., and J. Cohen-Cruz, eds. 1994. *Playing boal: Theatre, therapy, activism*. New York: Routledge.

Sharma, N. 2006. *Home economics: Nationalism in the making of "migrant workers" in Canada*. Toronto: University of Toronto Press.

Sharma, N. 2012. The "difference" that borders make: "Temporary foreign workers" and the social organization of unfreedom in Canada. In *Legislated inequality: Temporary labour migration in Canada*, eds. P. T. Lenard, and C. Straehle, 26–47. Montreal: McGill-Queen's University Press.

Sinding, C., R. Gray, and J. Nisker. 2008. Ethical issues and issues of ethics. In *Handbook of the arts in qualitative research*, eds. J. G. Knowles, and A. L. Cole, 459–468. Los Angeles: Sage Publications.

Skrivankova, K. 2010. *Between decent work and forced labour: Examining the continuum of exploitation*. New York: Joseph Rowntree Foundation.

Smith, A. 2016. Migration, development and security within racialised global capitalism: Refusing the balance game. *Third World Quarterly* 37(11): 2119–2138.

Strauss, K. 2012. Coerced, forced and unfree labour: Geographies of exploitation in contemporary labour markets. *Geography Compass* 6(3): 137–148.

Thomas, M. 2016. Producing and contesting "unfree labour" through the Seasonal Agricultural Worker Program. In *Unfree labour? Struggles of migrant and immigrant workers in Canada*, eds. A. Choudry, and A. Smith, 21–36. Oakland: PM Press.

Vosko, L. F. 2016. Blacklisting as a modality of deportability: Mexico's response to circular migrant agricultural workers' pursuit of collective bargaining rights in British Columbia, Canada. *Journal of Ethnic and Migration Studies* 42(8): 1371–1387.

Weiler, A., C. Levkoe, and C. Young. 2016. Cultivating equitable ground: Community-based participatory research to connect food movements with migrant farmworkers. *Journal of Agriculture, Food Systems, and Community Development* 6(2): 73–87.

J. Adam Perry (www.adamperry.ca) is a Postdoctoral Research Fellow cross appointed between the School of the Arts and the School Social Work at McMaster University in Hamilton, Canada. His research interests include migrant justice, migrant mobility, immigration policy, precarious work, employment policy, and community-based arts practice.